THE ILLUSTRATED DICTIONARY OF

The
HUMAN BODY

Design: Steven Hulbert and Jane Brett
Illustrations: Jeremy Gower and Matthew White (B.L. Kearley Ltd);
Oxford Illustrators Ltd; Jeremy Pyke.

Consultant: Alan M. Emond MA, MB, MD, MRCA, Consultant Paediatrician,
Bristol Royal Hospital for Children, Bristol, UK

Printed in Great Britain.

ISBN 1 85471 601 8

THE ILLUSTRATED DICTIONARY OF

The
HUMAN BODY

Contributors
Mal Sainsbury
Merilyn Holme
Jo Paker

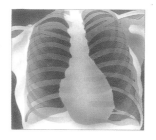

BLOOMSBURY

Reader's notes

The entries in this dictionary have several features to help you broaden your understanding of the word you are looking up.

- Each entry is introduced by its headword. All the headwords in the dictionary are arranged in alphabetical order.

- Each headword is followed by a part of speech to show whether the word is used as a noun, adjective, verb or prefix.

- Each entry begins with a sentence that uses the headword as its subject.

- Words that are bold in an entry are cross references. You can look them up in the dictionary to find out more information about the topic.

- The sentence in italics at the end of an entry helps you to see how the headword can be used.

- Many of the entries are supported by illustrations. The labels on the illustrations highlight the key points of information and will help you to understand some of the science behind the entries.

- Many of the labels on the illustrations have their own entries in the dictionary and can therefore be used as cross references.

abdomen   *noun*
The abdomen is the large area between the chest and the pelvis. It contains many important **organs**, such as the **stomach**, the **intestines**, the **spleen** and the **liver**. These organs are wrapped in a thin lining called the **peritoneum**. A large muscle called the **diaphragm** lies across the top of these organs and separates them from the **lungs**.
The front of the abdomen is made of layers of muscle.

abrasion ► graze

abscess   *noun*
An abscess is a collection of **pus**. It is a red, swollen lump. An abscess can develop at any place in the body. It builds up in tissue which is infected with **bacteria**. The pus in an abscess contains **white blood cells**. These fight the bacteria.
The gum abscess made her face swell.

absorption   *noun*
Absorption describes how body tissue takes in other substances. Food is broken down and absorbed in the **stomach** and **intestines** and nutrients carried in the bloodstream to all parts of the body.
The blood absorbs oxygen from the lungs.
absorb   *verb*

Achilles' tendon   *noun*
The Achilles' tendon is the thick **tendon** at the back of the ankle. It attaches the calf muscles to the bone of the heel.
His Achilles' tendon was damaged playing football.

acid   *noun*
An acid is a kind of chemical substance. The stomach produces acid which breaks down and helps **digest** food.
Stomach acid kills germs in food.

acne   *noun*
Acne is a skin disorder. It happens when the **sebaceous glands** in the skin become infected. The sebaceous glands produce an oil called **sebum**. This comes out onto the surface of the skin through the pores. If the pores are blocked, an infection builds up inside the sebaceous glands. Acne often starts when the **hormones** produced during puberty make the sebaceous glands produce more sebum.
He used a special ointment to treat his acne.

acupuncture   *noun*
Acupuncture is a way of preventing and treating illness and relieving pain. In acupuncture, fine needles are inserted into different points on the body. In China, acupuncture is sometimes used instead of an **anesthetic**, to block pain during an **operation**. The patient stays awake but seems to feel little or no pain.
Acupuncture can help to relieve back pain.

acute *adjective*
Acute describes an illness or a pain which suddenly develops or which suddenly becomes worse. The opposite of acute is **chronic**.
She felt acute pain in her ankle after she fell.

Adam's apple ► larynx

adenoids *noun*
The adenoids are glandular tissue which is found inside the throat at the back of the nose. The adenoids are similar to the **tonsils**. They catch and destroy **bacteria** entering the body when a person breathes. They also help the body to build up a resistance, or **immunity**, to **infection**. Adenoids that become infected and swollen can make breathing difficult.
Adenoids can be removed by a simple operation.

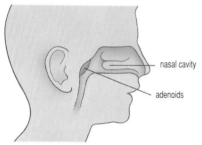
nasal cavity
adenoids

adolescence *noun*
Adolescence is the **age** between childhood and adulthood. Adolescence is a time of physical and emotional change. It takes place very approximately from 10 years old to 20 years old. At the beginning of adolescence, there is a period of rapid physical change and sexual development called **puberty**. Girls go through greater physical change than do boys. But adolescence is usually shorter for girls than for boys.
A boy's voice breaks during adolescence.

adrenal gland *noun*
The adrenal gland is a small **organ** that releases **hormones**. There are two adrenal glands, one on top of each **kidney**. The hormones they release help the body react to **stress** and allow the changes of **puberty** to begin.
The adrenal glands help to control the level of glucose in the bloodstream.

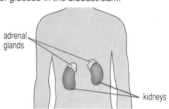
adrenal glands
kidneys

adrenalin *noun*
Adrenalin is a **hormone** produced by the **adrenal glands**. Adrenalin increases **heartbeat** and **respiration** in response to **stress**.
Adrenalin prepares the body for danger.

adult *noun*
An adult is a person who is fully grown and **mature**. Adults are sexually developed and can **reproduce**. People are adults once they have been through **adolescence**.
People spend the longest part of their life as an adult.

age *noun*
Age is a stage in a person's life. Childhood, **adolescence** and adulthood are all ages.
Old age is the last stage in a person's life.

aging *noun*
Aging is the process of becoming older. People grow during childhood and adolescence. Then they start gradually to age and their body slowly breaks down. A healthy **lifestyle** and medical treatment can slow down aging and many people can now live to more than 80 years old.
Aging can cause the joints to become stiff.

AIDS   *noun*
AIDS is a disease which breaks down the body's defence against illness, or **immune system**. AIDS stands for Acquired Immune Deficiency Syndrome and is caused by a **virus** called **HIV**. A person may have HIV in their body for many years before the actual disease of AIDS develops. AIDS results from damaged immune systems and the person cannot fight off diseases such as pneumonia.
There is no known cure for AIDS.

albino   *adjective*
Albino describes someone who is born without any colouring in their skin, hair or eyes. Colouring is usually caused by **melanin**, a brown pigment. An albino person's eyes are clear and may look pink because the blood vessels show through.
The albino child had pure white hair.
albinism   *noun*

alimentary canal   *noun*
The alimentary canal is a tube inside the body that runs from the **mouth** to the **rectum**. It includes the **esophagus**, **stomach** and **intestines**. Food is taken into the body through the mouth and passes along the alimentary canal to the **anus**.
The mouth is the top opening of the alimentary canal.

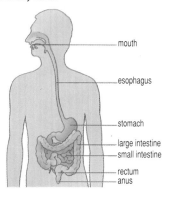

mouth

esophagus

stomach

large intestine
small intestine

rectum
anus

allergen   *noun*
An allergen is a substance which causes an **allergy**.
Grasses, cat fur and house dust are examples of allergens.

allergy   *noun*
An allergy is an unusual reaction to a substance. A substance that causes an allergic reaction is called an **allergen**. If a person with an allergy comes in contact with an allergen, they may develop a skin rash, a runny nose or itching eyes. **Hay fever** is an allergy to the pollen of flowers.
The boy has an allergy to cow's milk.
allergic   *adjective*

alternative medicine   *noun*
Alternative medicine describes treatments which do not rely on **drugs** and **surgery**. People who practise alternative medicine try to treat the whole patient. Alternative medicine is suitable for the treatment of many conditions.
Acupuncture and osteopathy are examples of alternative medicine.

alveoli   *noun*
Alveoli are the millions of tiny sacs at the end of the air passages in the lungs. The alveoli are surrounded by tiny blood vessels called **capillaries**. Oxygen in the air is breathed in during **respiration** and passes through the alveoli into the capillaries. Carbon dioxide passes from the **bloodstream** into the alveoli.
The alveoli are part of the respiratory system.

Alzheimer's disease   *noun*
Alzheimer's disease is a disorder of the **brain**. It affects people from the age of about 40 onwards. People with Alzheimer's disease suffer from loss of memory and confusion. They eventually become unable to look after themselves. There is no treatment and no cure for Alzheimer's disease.
Some scientists believe that Alzheimer's disease can be passed on by parents to their children.

amino acid   *noun*
An amino acid is a substance which makes up **proteins**. There are about 20 amino acids in the human body. They combine in different ways to make all the proteins needed. The human body can make about 10 amino acids. The other 10 are found in foods, such as cheese, meat and fish.
Amino acids are an essential part of a person's diet.

amnesia   *noun*
Amnesia is a loss of **memory**. A person might lose all their memory or may only lose their memory for certain events. Amnesia may happen after a disease or an injury to the **head**. It can also occur as a reaction to an emotional upset.
She suffered amnesia after the road accident.

amniotic fluid   *noun*
Amniotic fluid is the fluid inside the **uterus** of a pregnant woman. The **fetus** floats in the uterus, surrounded and protected by amniotic fluid.
Amniotic fluid lies inside the amniotic sac.

amniotic fluid

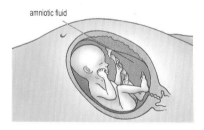

analgesic   *noun*
An analgesic is a **drug** which relieves pain, without making a person **unconscious**. A person with a **headache** might take an analgesic called aspirin.
The doctor prescribed a strong analgesic to ease his patient's aching back.

anatomy ► page 10

anemia   *noun*
Anemia is a disorder of the **blood**. Anemia happens when the blood has less **hemoglobin** than it should have. Anemia may result from a loss of blood after an accident or if the body cannot make enough red blood cells. People with anemia look pale and may feel tired.
The woman with anemia was treated with iron tablets.

anesthesia   *noun*
Anesthesia is the loss of feeling in all or part of the body. It can be produced by **drugs**, called **anesthetics**. Two forms of alternative medicine, hypnosis and **acupuncture**, can be used to cause anesthesia. It can also sometimes happen as a result of disease or injury, especially to the **nervous system**.
The boy needed anesthesia while his leg was mended.

anesthetic   *noun*
An anesthetic is a **drug** which causes a loss of feeling in the body. This is known as **anesthesia**. There are two main kinds of anesthetic. General anesthetics are used during surgical **operations**. They make the patient **unconscious** and unable to feel any pain. Local anesthetics cause a loss of feeling in only part of the body.
The dentist gave her a local anesthetic before pulling out the tooth.

anger ► feelings

angina   *noun*
Angina usually describes **spasms** of **pain** which people feel in their chest when too little oxygen reaches the **heart**. This can happen when the **blood vessels** leading to the heart muscle, called the **coronary arteries**, become partly blocked with fatty material carried in the blood.
Overweight people who smoke can suffer angina.

ankle   *noun*
The ankle is a part of the body that lies
between the leg and the foot. It is made up of
two bony lumps, one on either side of the
leg. These are the ends of the two lower leg
bones, called the **tibia** and the **fibula**. The
Achilles' tendon lies at the back of the
ankle. It helps to move the ankle **joint**.
*The ice-skater's strong ankles helped him to
perform daring leaps and turns.*

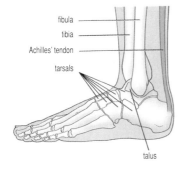

fibula
tibia
Achilles' tendon
tarsals
talus

anorexia   *noun*
Anorexia is a loss of **appetite**. A person with
anorexia does not want to eat anything.
Anorexia nervosa is an emotional illness
suffered by some **adolescent** girls and
young women. They refuse to eat and suffer
a very severe loss of weight.
*She was very thin because she was suffering
from anorexia.*

anterior   *adjective*
Anterior describes the parts of the human
body that face forwards. The opposite of
anterior is **posterior**.
The chest is an anterior part of the body.

anti-   *prefix*
Anti- is a prefix meaning against or
preventing.
*The traveller took a course of anti-malarial
tablets to protect him from the disease.*

antibiotic   *noun*
An antibiotic is a kind of **drug** which fights
bacteria in the human body. Scientists have
developed many kinds of antibiotic to fight
the different bacteria that cause **diseases**.
Some antibiotics are made from natural
substances, such as fungi or moulds.
Penicillin is one of these. Other antibiotics
are made from synthetic materials.
*A course of antibiotics cured his throat
infection.*

antibody   *noun*
An antibody is a part of the body's defence,
called the **immune system**. When **bacteria**
or **viruses** enter the body, they produce
substances called **antigens**. Special **white
blood cells** called lymphocytes react to the
presence of antigens and make antibodies.
The antibodies attack and kill the invading
bacteria or viruses.
Antibodies help a person to fight infection.

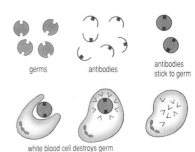

germs   antibodies   antibodies
                                    stick to germ

white blood cell destroys germ

antidote   *noun*
An antidote is a substance that fights the
harmful effects of poison in the body. Most
poisons have only one kind of antidote. If the
wrong kind of antidote is used with a poison,
it may make the effects of the poison worse.
*When he was bitten by a poisonous snake,
the antidote stopped the farmer becoming
seriously ill.*

9

anatomy *noun*

Anatomy is the study of how the body is put together. It involves looking at the different parts of the body, grouping them together in a **system**. For example, the anatomy of the digestive system includes all the parts of the body concerned with the digestion.

She cut open the heart to study its anatomy.

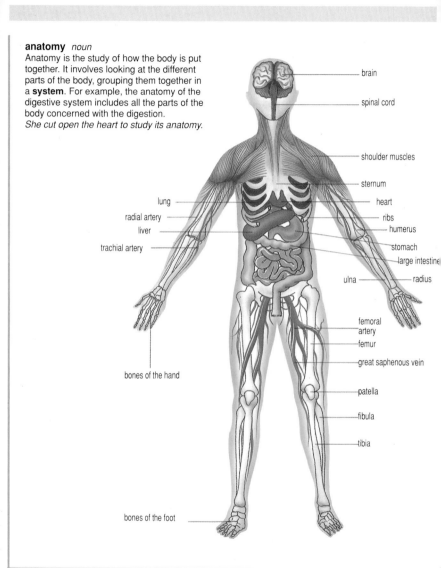

brain

spinal cord

shoulder muscles

sternum

lung

radial artery

liver

trachial artery

heart

ribs

humerus

stomach

large intestine

ulna

radius

femoral artery

femur

great saphenous vein

bones of the hand

patella

fibula

tibia

bones of the foot

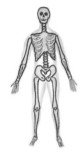

The skeletal system includes the body's framework of bones.

The muscular system includes all the muscles that help the body to move.

The digestive system involves all the parts of the body that process food.

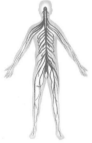

The nervous system connects nerve cells in the body with the brain.

The respiratory system consists of the lungs, nose and air passages.

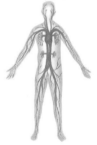

The circulatory system includes the heart and all the blood vessels.

The kidneys, ureters and bladder make up the urinary system.

The endocrine system consists of glands which produce hormones.

The reproductive system includes all the organs to do with sex.

antigen *noun*
An antigen is a kind of **protein** which is made when **germs**, such as **bacteria** and **viruses**, have found their way inside the human body. Special **blood cells** notice the antigens in the bloodstream, go into action and produce **antibodies** to fight and kill the germs.
Antigens warn the body's defence system that germs have invaded the body.

antiseptic *noun*
An antiseptic is a substance which is used to stop **infection**. It kills **germs** and prevents them spreading. Antiseptics are put on skin and mucous membrane to help stop **infection**. **Disinfectants** are used to kill germs on non-living things.
She put antiseptic on the cut to stop it becoming infected.

antitoxin *noun*
An antitoxin is a kind of **antibody** that is part of the body's defence, or **immune system**. Antitoxins are made by the **blood** to protect it from a poison, or **toxin**, which has entered the body.
Antitoxins fight the poisons produced by some bacteria.

anus *noun*
The anus is an opening in the body. It lies at the end of the **rectum** and is the final part of the alimentary canal. Solid waste material is produced during the **digestion** of food and passes out of the body through the anus.
The anus is the lower opening of the digestive system.
anal *adjective*

anvil *noun*
The anvil is one of three tiny bones inside the **ear**, which are known as the **ossicles**. The anvil joins the **hammer** to the **stirrup**. The three bones carry sound vibrations from the **eardrum** to the **inner ear**.
The anvil is a bone which passes on sound in the process of hearing.

anxiety ► **feelings**

aorta *noun*
The aorta is one of a number of tubes which carry **blood** around the body. These tubes are known as **arteries**. The aorta is the largest artery in the body. **Blood** flows along the aorta from the **heart** to join with other arteries which take the blood to all parts of the body.
The aorta carries blood from the heart to the rest of the body.

appendicitis *noun*
Appendicitis is an **infection** of a part of the body called the **appendix**. The infection is caused by **bacteria** and makes the appendix become inflamed. It swells up and fills with **pus**. A person with appendicitis feels pain in the lower right-hand corner of the **abdomen**, and has a slight fever. Usually, the infected appendix has to be removed in hospital.
The boy's appendicitis was cured when the surgeon took his appendix out.

appendix *noun*
The appendix is a part of the body joined to the first part of the large intestine, called the **cecum**. It is worm-shaped and about eight centimetres long. The appendix can become infected and cause **appendicitis**.
The appendix lies in the lower right side of the abdomen.

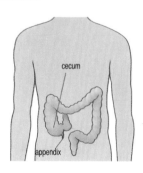

appetite   *noun*
Appetite is the natural, healthy desire for food. People may lose their appetite when they are ill. Some strong, unpleasant **feelings**, such as anxiety, can cause a loss of appetite.
He had such a good appetite that he ate three eggs for breakfast.

arm   *noun*
An arm is one of the upper **limbs** of the body. It is made up of the upper arm, the **elbow** and the forearm. The arm is joined to the body at the **shoulder** and to the **hand** at the **wrist**. The upper arm bone is called the **humerus**. The two forearm bones are the **radius** and the **ulna**. **Muscles** attached to the bones help the arm to move.
The mother carried the sleeping baby in her arms.

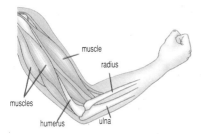

muscle
radius
muscles
humerus
ulna

artery   *noun*
An artery is a flexible tube inside the body. It has thick, strong walls. Arteries carry blood mixed with oxygen from the **heart** to all parts of the body. Tiny **muscles** in the artery walls help the heart to push the blood along. This flow of blood along the arteries can be felt as the **pulse**. The walls of arteries may become thicker and harder, and narrowed by fat deposits. This makes it more difficult for blood to flow through and can cause **heart disease**.
The carotid arteries lie in the neck.

arthritis   *noun*
Arthritis is a disorder of the **joints**, in which the joints become painful, stiff and swollen. It may be caused by **inflammation** or **infection** in the joint. Sometimes it is a **symptom** of another **disease**. Some people have arthritis in old age when their joints begin wearing out. There is no cure for arthritis, but **drugs** can help to ease the pain and inflammation.
In some kinds of arthritis, diseased joints can be replaced with artificial ones.

artificial   *adjective*
Artificial describes something which is made by people and is not natural. A person who has lost a **limb**, such as a leg, could have an artificial leg fitted. Artificial **respiration** is used to save the life of a person who has stopped breathing. It forces air into the lungs and helps the person to start breathing again naturally.
The children learned how to do artificial respiration.

asthma   *noun*
Asthma is a breathing disorder. It happens when small tubes in the **lungs** become narrowed. This makes it difficult for a person to breathe. The chest feels tight and the person makes a wheezing sound. Asthma is usually caused by an **allergy**. **Infection**, cold weather, tiredness or **stress** can also bring on asthma. **Drugs** can help people who suffer from asthma attacks.
His asthma attack was relieved by using an inhaler.

astigmatism   *noun*
Astigmatism is a disorder of the **eye** which results in blurred, or distorted, **sight**. It happens when the outer layer of the eye, called the **cornea**, curves the wrong way. It also happens if the part of the eye called the **lens** is slightly out of shape. If it is serious, astigmatism can be corrected by wearing spectacles.
Almost everyone has a small amount of astigmatism.

athlete's foot   *noun*
Athlete's foot is a skin infection, which often occurs between the **toes**. It is caused by one of two kinds of **fungus**. The skin becomes blistered, sore and itchy. If the **blisters** break, **bacteria** may enter the skin and cause another infection. Athlete's foot can be treated with lotions, ointments or powders bought at a **pharmacy**.
His sister caught athlete's foot at the swimming pool.

atlas   *noun*
The atlas is a ring-shaped **bone**. It is the part of the **spine** that supports the **skull**. The atlas is the first of the bones in the neck, or **cervical vertebrae**.
The atlas is joined to a vertebra called the axis.

atrium (plural **atria**)   *noun*
Atrium is a word which means a hollow, or **cavity**. It is used to refer to the two upper cavities in the **heart** where blood is collected during **circulation**. They are called the left atrium and the right atrium. The atria are lined with strong **muscles**. Another name for atrium is auricle.
The right atrium receives blood from the body and the left atrium receives blood from the lungs.

audible   *adjective*
Audible describes a sound that can be heard.
The explosion was audible from a great distance.

auditory   *adjective*
Auditory describes anything to do with **hearing**. The auditory **nerve** is the main nerve of the **inner ear**. The auditory range is the total number of sounds that a person can hear.
A fly got stuck in her auditory canal.

aural   *adjective*
Aural describes anything to do with the **ear**. Aural **surgery** involves operations on the ear.
The doctor used an aural instrument to examine her ear.

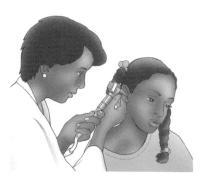

auricle ▶ **atrium**

autism   *noun*
Autism is a rare disorder of the **mind**. It usually appears in childhood. Children with autism are unable to communicate with other people. They often cannot speak and may injure themselves on purpose. Children with autism may be mentally handicapped but they are physically normal. It is not known what causes autism. The **symptoms** can be treated with a special kind of teaching, or behaviour modification.
Autism affects boys more often than girls and it cannot be cured.
autistic   *adjective*

14

axilla (plural **axillae**)   *noun*
Axilla is the scientific term for an armpit. It is a hollow, or **cavity**, below the shoulder joint. Strong **muscles** make up the front and back walls of the axillae. These muscles are attached to the chest and shoulder blade and help to move the **arms**.
An axilla contains sweat glands and lymph glands.
axillary   adjective

axis   *noun*
The axis is a ring-shaped **bone**. It is a part of the **spine**. The axis is the second of the bones in the neck, called the **cervical vertebrae**.
The axis is one of seven bones in the neck.

baby   *noun*
A baby is a human child, from birth to the age of about one year. Unlike the young of some other mammals, human babies are completely helpless at birth.
The baby was just beginning to walk at the age of eleven months.

back   *noun*
The back is the part of the human body between the **neck** and the **pelvis**. It is a **posterior** part, which means it faces backwards. The back includes the **spine**, **shoulders**, shoulder blades and **ribs**.
The walker carried a rucksack on his back.

backache   *noun*
Backache is a very common kind of **pain**. It can be felt in any part of the area around the **spine**. Backache has many different causes. The most usual ones are strained **muscles** or **ligaments**, and disorders of the **bones** or **nerves** in the back. Painkilling drugs, called **analgesics**, can ease the pain of backache. Other **treatments** depend on the cause of the backache.
He had backache after lifting the heavy box.

15

backbone ► **spine**

bacteria (singular **bacterium**)   *plural noun*
Bacteria are tiny living things. They are a
kind of **germ**. Most bacteria are harmless,
but some can cause **diseases** when they
invade the body. They are so small they
cannot be seen without a microscope.
*Bacteria in the drinking water made the
people at the campsite ill.*

balance   *noun*
Balance is the ability to stand or move
upright and not fall over. It is controlled by
the part of the brain called the **cerebellum**,
which is helped by the **inner ear**. When the
position of the body changes, the cerebellum
sends messages to the **muscles** telling them
to keep the body in balance. Upset balance,
or **dizziness**, can be caused by travel
sickness, ear infections or when not enough
blood and oxygen reach the brain.
*The boy found it hard to keep his balance the
first time he used a skateboard.*

ball and socket joint ► **joint**

behaviour   *noun*
Behaviour is the way people act or react.
Some kinds of behaviour, such as **breathing**,
do not have to be learned. They are
instinctive. Other kinds of behaviour, such
as speaking, have to be learned.
*Smiling is a kind of behaviour that babies
learn when they are a few weeks old.*

benign   *adjective*
Benign describes something that is not
dangerous or life-threatening. A benign
disease is mild and does not last long. A
benign growth in the body, or **tumour**, does
not contain dangerous **cancer cells** that will
spread.
*She recovered quickly after the benign
tumour was removed.*

beriberi   *noun*
Beriberi is a **disease** which is caused by a
lack of vitamin B_1 in a person's **diet** over a
long period. Vitamin B_1 is found in foods such
as wheatgerm, liver, beans and brown rice.
Beriberi affects the **muscles** in a person's
body, causing stiffness and pain. After a time,
it affects the **nervous system** and the
person may become paralysed. Beriberi is
treated with a series of vitamin B_1 **injections**,
and a diet rich in this vitamin.
*Beriberi used to be common in Asia when
white rice was sometimes the only food
available.*

biceps   *noun*
A biceps is a **muscle**. In the upper **arm**, the
biceps is the muscle at the front. It works
together with the **triceps**, to bend the arm at
the **elbow**. In the **thigh**, the biceps femoris
lies at the back and helps to bend the **knee**.
*The biceps is a flexor muscle, which means it
pulls the arm to bend at the elbow.*

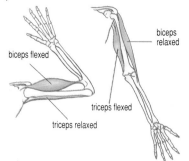

biceps relaxed
biceps flexed
triceps flexed
triceps relaxed

bile *noun*
Bile is a yellowish-green liquid which is made by the **liver**. Bile is made up of **cholesterol**, bile salts, **proteins** and urea, which is found in **urine**. It is stored in the **gall bladder**. It passes into the **intestine**, where it helps to break down, or **digest**, fats.
Bile is coloured by a substance called bilirubin.
bilious *adjective*

bilharzia *noun*
Bilharzia is a disease. It is caused by a worm that infects the **liver**, **kidneys**, **lungs** and other **organs**. The worms that cause the disease live in water in tropical countries. A person with bilharzia has **fever**, **cough**, muscle pain and irritation of the **skin**. There may also be internal bleeding. Bilharzia can be treated with a special **drug**. Bilharzia is also known as schistosomiasis.
He caught bilharzia from swimming in an infected lake.

binocular vision ► **sight**

biological clock *noun*
Biological clock is a term which describes the way the body's activities are timed. It controls the need to sleep and the need to wake up. It controls a person's body **temperature** and tells them when to eat or drink. A biological clock adapts to the pattern of day and night. A person who travels by aeroplane to a part of the world where this pattern is different, will find that their biological clock is upset for a few days.
Her biological clock told her it was lunchtime.

biology *noun*
Biology is the study of living things. It looks at how plants and animals are made, how they develop and behave, and where they live. A person who studies biology is called a biologist.
He learned about the circulation of the blood in the biology lesson.
biological *adjective*

biopsy *noun*
A biopsy is a kind of medical test. In a biopsy, a tiny piece of **tissue** is removed from the body. The tissue is then looked at under a microscope to see if there is any **disease** in the tissue.
The surgeon performed a biopsy to find out if the lump in her leg was harmful.

birth ► **childbirth**

bite *noun*
A bite is a wound on the **skin** made by an animal or an insect. Most bites are not dangerous. They only need to be kept clean, so they do not become infected. But poisonous snakes and spiders inject **toxins** into the body which must be treated with special drugs called **antidotes**. Some insects and other animals carry **diseases**, such as **malaria**, which they pass on to people when they bite them.
The midge bites on her arm itched for a few days.

bladder *noun*
The bladder is a strong, muscular bag inside the body. It lies just behind the bone called the **pubis**. Liquid waste, called **urine**, is produced by the **kidneys** and stored in the bladder. When the bladder is full, it pushes the urine out of the body through a tube called the **urethra**.
The bladder of an adult can hold about half a litre of urine.

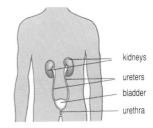

kidneys

ureters

bladder

urethra

bleed   *verb*

Bleed describes the action of blood when it flows from any part of the body. When the **skin** is cut, the wound bleeds. Some kinds of injury or disease can cause inner body parts to bleed. This is called internal bleeding.
His knee bled after he fell on the stones.

blindness   *noun*

Blindness is the lack of **sight** in one or both **eyes**. It may be present at birth, or it may happen at any time in a person's life. Blindness happens as a result of a disorder of the eye itself, such as **glaucoma** or **cataract**. It can also happen if there is something wrong with the part of the brain that controls sight. Blind people can have special teaching to help them to lead a normal life.
People suffering from blindness can learn to read using a system of raised dots, called braille.

blink   *verb*

Blink describes the rapid, up and down movement of the **eyelids**. People blink their eyelids every few seconds to keep their eyes moistened with tears. They blink to protect their eyes from very bright light or smoke in the air. They also blink as a **reflex** action when an object suddenly comes near their eyes.
People blink their eyelids without thinking about it.

blister   *noun*

A blister is a bump on the skin. It contains a clear, watery fluid called **serum**. Blisters usually form when the skin is irritated, rubbed or pinched. A **burn** may cause a blister. A blood blister is caused by burst **blood vessels** under the skin.
She had a blister on her heel because her shoes were too tight.

blood   ► page 20

blood cell   *noun*

A blood cell is a part of **blood**. There are three main kinds of blood cell. Red blood cells carry **oxygen** to all parts of the body. White blood cells help the body to fight **infection** and **disease**. **Platelets** stop leaks in blood vessels and help the blood to clot.
Blood cells are also known as corpuscles.

blood count ► blood test

blood group   *noun*

A blood group is a type of **blood**. Scientists have sorted human blood into different blood groups. A person's blood group depends on what kinds of **protein** are present in their blood. The most common method of sorting blood is called the ABO system. Under this system, the blood groups are A, B, O or AB. When a person is given blood during a **transfusion**, it is essential that this blood type matches their own.
The test showed that her blood group was B.

A and O are the most common blood types

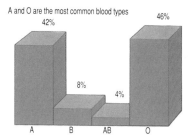

blood pressure   *noun*
Blood pressure is the pushing, or pressure, of the blood against the walls of the **arteries**. The amount of blood pressure depends on how strongly and quickly the heart contracts and pushes the blood along the arteries. It also depends on how much blood is flowing and how elastic the artery walls are. Blood pressure can be measured with a special instrument.
High blood pressure can be a sign of heart disease.

blood test   *noun*
A blood test is really a series of tests that is done on a sample of **blood**. In a blood test, a very small amount of blood is taken from the body. This sample is examined under a microscope to see if the person is in good health. A blood test usually includes a count of the different kinds of **blood cell**.
The blood test showed she had a low count of red blood cells and was anemic.

blood vessel   *noun*
A blood vessel is a tube in the body through which blood flows, or **circulates**. The main blood vessels are the **arteries**, **veins** and **capillaries**.
The smallest blood vessels are the capillaries.

bloodstream   *noun*
The bloodstream is the flow of **blood** around the body.
Oxygen enters the bloodstream in the lungs and is carried round the body.

blushing   *noun*
Blushing is a sudden reddening of the skin of the face and neck. It happens when the tiny **blood vessels** near the surface of the skin become more open, or dilated. Extra blood rushes into the dilated vessels and makes the skin look red. Blushing may happen when a person becomes hot, or if they feel an emotion such as anger or excitement.
Blushing is most often linked to a feeling of embarrassment.

body temperature ► **temperature**

boil   *noun*
A boil is a skin disorder. It is an **infection** of a hair root or of a **sweat gland**. The infection is caused by **bacteria**. A boil begins as a red, painful lump on the skin. It grows bigger and after about three days a yellow head forms. The boil must burst and let out **pus** before it can heal. **Antibiotics** are sometimes used to treat boils.
Boils may come back if the skin is not kept clean.

bone   *noun*
Bone is the hard, stiff substance that makes up the body's framework, or **skeleton**. It supports the body and protects the delicate inner **organs**. Bone is mostly made up of **calcium** salts, held together with strong fibres. It is usually hollow and bone **marrow** and **minerals** are stored inside. There are two kinds of bone. **Compact bone** is hard and solid. **Cancellous bone** looks spongy. Humans have about 200 long and short bones.
The child's broken arm bone mended quickly.

bone marrow   *noun*
Bone marrow is a kind of soft tissue that is found in the hollow centre of **bones**. Red bone marrow has special **cells** that make fresh **blood** for the body. Yellow bone marrow is mostly **fat**.
Bone marrow produces over 100 million red blood cells every minute.

blood *noun*

Blood is a red fluid that continually flows through the **blood vessels**. Blood is pumped round the body by the **heart**. The blood transports food and oxygen to all living **cells** in the body, and takes away waste matter from them.

An adult male has about five litres of blood in his body.

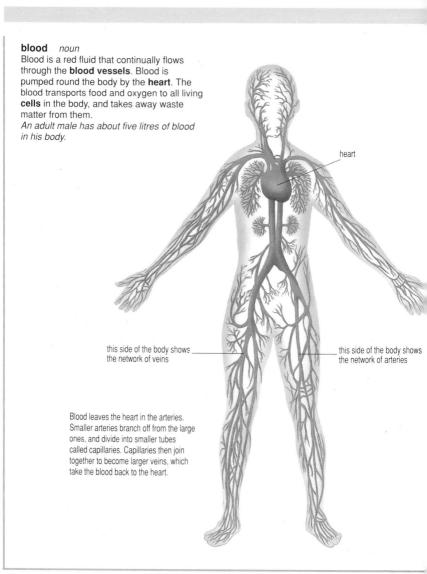

heart

this side of the body shows the network of veins

this side of the body shows the network of arteries

Blood leaves the heart in the arteries. Smaller arteries branch off from the large ones, and divide into smaller tubes called capillaries. Capillaries then join together to become larger veins, which take the blood back to the heart.

Blood consists mainly of a watery liquid called plasma. Floating in the plasma are red cells, white cells and platelets. The disc-shaped red blood cells take oxygen from the lungs to the tissues. White blood cells are larger than red blood cells. There are 500 red cells to every white one. White blood cells help to fight infection by attacking harmful bacteria. Small particles called platelets are important in blood clotting.

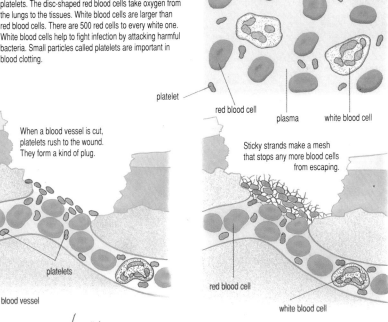

platelet

red blood cell

plasma

white blood cell

When a blood vessel is cut, platelets rush to the wound. They form a kind of plug.

platelets

blood vessel

Sticky strands make a mesh that stops any more blood cells from escaping.

red blood cell

white blood cell

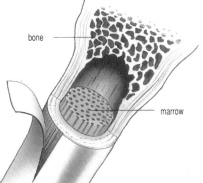

bone

marrow

Blood is made in the bone marrow. A red blood cell lives for about 120 days after it enters the bloodstream. When blood cells die, they are disposed of by the liver and the spleen. The bone marrow continuously makes new blood cells to replace dead cells.

botulism *noun*
Botulism is a rare kind of food poisoning. The **toxin** that causes botulism is usually found in foods that have not been properly prepared for canning or preserving. Botulism attacks the **nervous system**. The first symptoms include general weakness, blurred vision, **nausea** and **diarrhoea**. Botulism needs immediate treatment with **antitoxins** as it develops very quickly and is often fatal.
Modern methods of canning food have made botulism very rare.

bowel ► **intestine**

brain ► page 24

brain stem *noun*
The brain stem is the stalk-like base of the **brain**. It joins the **cerebrum** to the **spinal cord**. The brain stem controls the **involuntary systems**, such as breathing, heartbeat and blood circulation.
Your brain stem keeps your heart beating regularly.

breast *noun*
The breast is the front of the chest. In **adolescent** girls and women, the breasts are the **mammary glands**. Girls begin to grow breasts at **puberty**. When fully grown, each breast contains 15 to 20 milk glands that produce milk after **childbirth**. Each of these glands has a tube, or **duct**, that leads to the **nipple**. The rest of the breast is fatty **tissue**.
The baby drank milk from his mother's breast.

breast-feeding *noun*
Breast-feeding is the act of feeding a **baby** at the **breast**. During the first two or three days of breast-feeding, the breasts produce a substance called **colostrum**. Milk begins to flow between three and five days after birth. Breast milk is easy to **digest** and protects the baby against **allergy** and **disease**.
Breast-feeding helps the mother to form a close bond with her baby.

breathing ► **respiration**

breathing rate *noun*
The breathing rate is the speed at which air is drawn into and let out of the **lungs**. It varies according to how active or still a person is. The more active a person is, the faster their breathing rate becomes.
The sleeping child's breathing rate was slow and even.

bronchial *adjective*
Bronchial describes anything to do with the air passages called the **bronchi**. A bronchial dilator, or bronchodilator, is a kind of drug that widens the airways to the lungs to help people with breathing difficulties.
Bronchial disorders affect the lungs and can cause breathing problems.

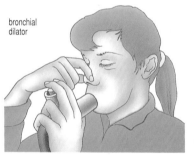

bronchial dilator

bronchiole *noun*
A bronchiole is a tiny tube in the **lung**. As it reaches the lungs, the windpipe, or **trachea**, branches into two **bronchi**. In turn, the bronchi branch into tiny 'twigs' called bronchioles. A bronchiole is about half a millimetre in diameter and lined with smooth **muscle**. Air passes through the bronchioles into the **alveoli**. Bronchiolitis is inflammation of the bronchioles that sometimes affects babies and young children.
There are more than 250,000 bronchioles in a person's lungs.

bronchitis *noun*
Bronchitis is **inflammation** of the **bronchi**.
Acute bronchitis begins suddenly and ends
after a short course of treatment. It may
follow a nose or throat infection. **Chronic**
bronchitis develops slowly and is difficult to
treat.
*Her bronchitis improved when she stopped
smoking.*

broncho- *prefix*
Broncho- is a prefix meaning to do with the
bronchi. Bronchopneumonia is a kind of
pneumonia that affects the bronchi.
*He had a kind of X-ray called a bronchogram
to see inside his lungs.*

bronchoscope *noun*
A bronchoscope is a kind of instrument. It is
a long, thin tube, with a light on the end.
Doctors use bronchoscopes to look down a
patient's **trachea** and **bronchi**. They can
also be used for removing objects that have
been accidentally breathed in.
*He looked down the bronchoscope and saw
a button in the little boy's windpipe.*

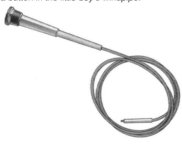

bronchus (plural **bronchi**) *noun*
The bronchi are two pipes that branch off
from the **trachea**. One bronchus goes to
each **lung**. The bronchi carry air to and from
the lungs, during the process of **respiration**.
They are lined with special **cells** that act like
hairs, sweeping **mucus** and dust away from
the lungs towards the throat.
Bronchitis is inflammation of the bronchi.

bruise *noun*
A bruise is a minor injury caused by a blow to
the body or a fall. It appears beneath the skin
as a purplish mark. The colour of a bruise is
caused by blood escaping from tiny **blood
vessels** that are broken by the sudden
pressure on the skin. The colour fades
gradually. It changes to blue, then brown,
then yellow, before it disappears. A bruise
may be painful and swollen.
*The pain and swelling of a bruise can be
helped by applying an ice pack.*

bulimia *noun*
Bulimia is an eating disorder. People with
bulimia have frequent periods of overeating,
called binges, which they cannot control.
After each binge, they usually make
themselves sick so as not to gain weight.
Bulimia is an emotional illness that is often
connected to **anorexia** nervosa.
*Bulimia may be treated with psychotherapy,
or antidepressant drugs.*

bunion *noun*
A bunion is a hard swelling. It occurs on the
big toe **joint** at the side of the foot. The
tissues over the swelling become red,
inflamed and often painful. A bunion is made
worse by tight-fitting shoes. The disorder is
often **hereditary**, which means it is passed
on in families. If a bunion becomes very
severe, it can be cured by an **operation**.
*Wearing loose shoes helped to ease the
discomfort of her bunion.*

23

brain *noun*

The brain is an **organ** in the body. It is the control centre of the body's **nervous system.** The brain sends out and receives hundreds of messages every second. It controls every activity of the body, such as moving the **muscles**, and **breathing**. It also controls all the functions of the **mind**, such as **thinking** and remembering. The brain is made up of soft tissue which is protected by the hard bones of the **skull**.

A special part of the brain controls every function of the body.

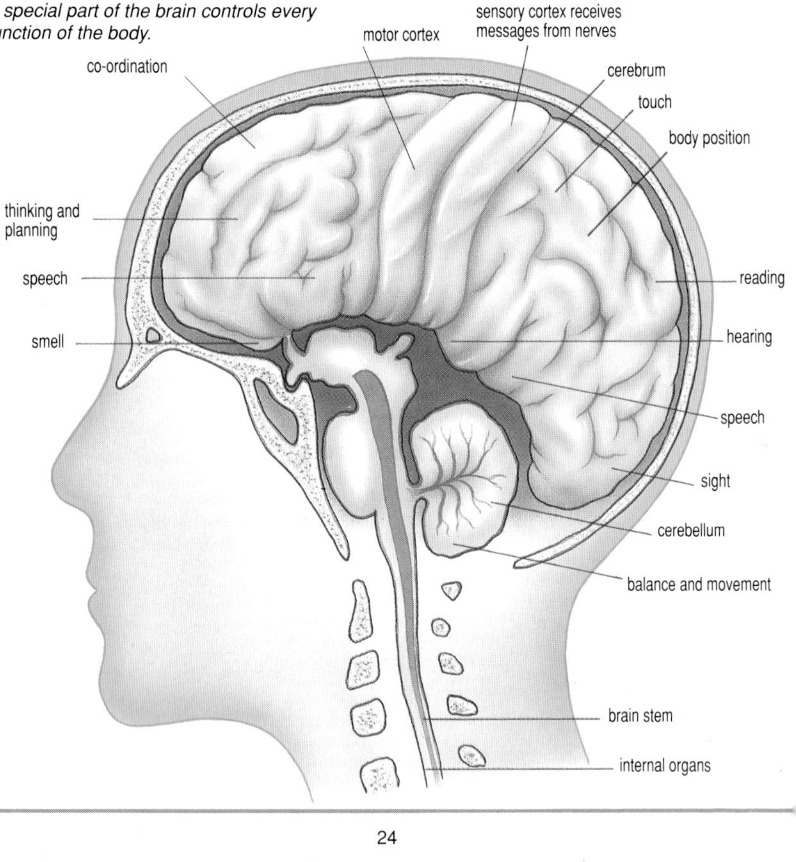

co-ordination

motor cortex

sensory cortex receives messages from nerves

cerebrum

touch

body position

thinking and planning

speech

smell

reading

hearing

speech

sight

cerebellum

balance and movement

brain stem

internal organs

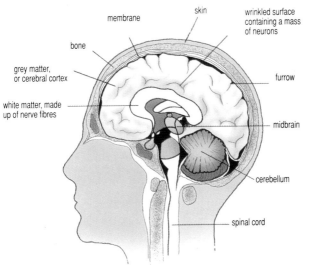

membrane

skin

wrinkled surface containing a mass of neurons

bone

grey matter, or cerebral cortex

furrow

white matter, made up of nerve fibres

midbrain

cerebellum

spinal cord

left cerebral hemisphere

right cerebral hemisphere

Messages travel up nerve pathways to the thalamus. The thalamus co-ordinates messages to and from the brain. As they go, the pathways cross from one side to the other. So the left cerebral hemisphere is linked with the right side of the body, and the right cerebral hemisphere controls the left side of the body. The pathways from the two sides of the cerebellum also cross over.

thalami

right cerebellar hemisphere

left cerebellar hemisphere

burn   *noun*

A burn is damage to the **skin** caused by
heat, cold, radiation or certain chemicals.
Burns are grouped according to how deep
they are. First-degree burns produce a patch
of red skin and heal without forming a **scar**.
Second-degree burns go deeper and cause
the skin to blister. Third-degree burns destroy
the skin and the **tissues** underneath, such
as fat or muscle. They leave an open area
and a skin **graft** is needed to prevent serious
scarring. People with large, deep burns
should have emergency medical help.
Sunburn is an example of a first-degree burn.

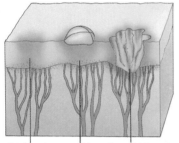

first degree burn second degree burn third degree burn

calcaneus ► **heel**

calcium   *noun*

Calcium is a mineral. The body needs
calcium for good health. It helps muscles to
contract and it also helps blood to clot
properly. Too little calcium in a body can lead
to bone diseases, such as **rickets**. Calcium
is found in foods such as milk and cheese.
*There was calcium in the milk that the
children drank.*

calorie   *noun*

A calorie is a unit of measurement. Calories
measure the **energy** value of food. Different
foods give different amounts of calories when
eaten and digested. The amount of calories a
person needs each day depends on their
age, sex, build and level of activity. A person
who eats too much and takes in too many
calories will put on weight.
*She went on a diet to reduce her calorie
intake and lose weight.*

canal   *noun*

A canal is a tube that carries food, liquid or
air from one part of the body to another.
A baby comes out through the birth canal.

cancer   *noun*

Cancer is a **disease** in which **cells** grow
without control. This cell growth can destroy
healthy **tissue** and endanger life. Cancers can
happen in any part of the body. Many kinds
can be completely cured, but cancer remains
a major cause of death in many countries.
*Many people try to avoid substances, such
as tobacco smoke, that can cause cancer.*

canine ► tooth

capillary *noun*

A capillary is the smallest type of **blood vessel**. Capillaries connect the smallest **arteries** with the smallest **veins**. They can be so tiny that only one blood cell at a time is able to pass along them. Capillaries carry blood rich in **oxygen** to the tissues, and carry away **waste** products.
There are capillaries in every part of the body.

carbohydrate *noun*

A carbohydrate is a source of energy. Carbohydrates are essential to the body. Too few carbohydrates can cause **fatigue** and depression. They should be eaten as part of a balanced diet. Too many can lead to obesity and **heart disease**. All sugars and starches are carbohydrates. Foods with a high carbohydrate content include potatoes, bananas, cereals and bread.
His diet was so high in carbohydrates that he grew fat.

carbon dioxide *noun*

Carbon dioxide is a gas. It is colourless and odourless. Carbon dioxide is produced in body **tissues** as a **waste** product. It dissolves in **blood**, and is carried by the blood to the **lungs**. Here it is passed into the **alveoli**, and is breathed out of the body.
Carbon dioxide is carried in the blood to the lungs.

carcinogen *noun*

A carcinogen is a word used for any substance known to cause **cancer**. Tobacco and asbestos are both carcinogens.
The scientists discovered that the substance was a carcinogen and warned people not to use it.

carcinoma *noun*

A carcinoma.is a type of cancer. Carcinomas form in the layers of tissue that make up the inside and the outside of most **organs** of the body. The most common carcinomas happen in the skin, stomach, lungs, prostate gland and breast. The growths are **malignant**, and can spread to other parts of the body. Carcinomas are usually treated by being surgically removed.
The surgeon successfully removed the carcinoma from the man's stomach.

cardiac *adjective*

Cardiac describes anything to do with the heart. Cardiac arrest is the sudden stopping of the heart.
Cardiac muscle is the medical name for the heart muscle.

cardiovascular system *noun*

The cardiovascular system includes the **heart** and the **blood vessels**, and circulates **blood** throughout the body. Blood is pumped by the heart through the **arteries**, **veins** and **capillaries**. The blood takes food and fuel to every living cell, and carries away waste products. There are 60,000 kilometres of tubes in the adult cardiovascular system, and about 5 litres of blood.
The heart pumps blood round the cardiovascular system.

caries *noun*

Caries describes the decaying of teeth and bones. Caries often involves **inflammation** of the surrounding soft **tissue**. It is most often used to describe tooth decay.
Caries had made a hole in his tooth which the dentist filled.

carotid artery ► artery

cartilage   *noun*
Cartilage is very strong **connective tissue**. It has a whitish colour or it is semi-transparent. Part of the **nose** is formed from cartilage, and so are some **ribs**. Cartilage is also found between some of the **vertebrae** of the spine, and as a covering on the surface of **joints**. Cartilage can cope with great pressures and tensions.
He tore the cartilage in his knee, and suffered great pain.

cataract   *noun*
A cataract is a non-transparent, or opaque, area in the **lens** of the **eye**. A cataract will cause a person to lose their sight gradually. Cataracts are quite painless and can be removed by surgery. Injury to the fetus may cause it to be born with cataracts. Cataracts may also form after injury to the eye or in later life. They may be a symptom of diseases such as **diabetes**.
The old woman had cataracts in her eyes and could not see.

catarrh   *noun*
Catarrh is a mucous discharge from the upper airway. It is a symptom of various **infections**, such as colds and hay fever. The **mucous membranes** lining the nose and throat become inflamed, producing large amounts of **mucus**.
He always suffered from catarrh when he caught a cold.

cauterize   *verb*
To cauterize is to destroy tissue on purpose by burning it. A doctor may use heat, cold, electricity or lasers to cauterize. Wounds that are likely to become infected can be cauterized, and so can lumpy scar tissue. Surgeons sometimes cauterize small blood vessels to stop them bleeding during operations.
He cauterized the wound to prevent infection.
cautery   *noun*

cavity   *noun*
A cavity is an enclosed space inside the body. It may be a large space, such as the abdominal cavity. It may be a small hole in a tooth caused by decay, or **caries**.
The oral cavity is the space inside the mouth.

cecum   *noun*
The cecum is the first part of the **large intestine**. It is found just beyond the point where the lower part of the **small intestine**, or **ileum**, joins the large intestine.
The appendix is attached to the cecum.

cell ► page 29

cerebellum   *noun*
The cerebellum is one of the three main parts of the **brain**. It is found just behind and above the **brain stem**. The cerebellum co-ordinates body movements, balance and posture.
She found it hard to balance because of damage to her cerebellum.

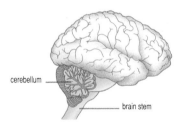

cerebellum

brain stem

cerebral cortex   *noun*
The cerebral cortex is part of the **brain**. It forms the outer layer of each part of the **cerebrum** and has a grey colour. The cerebral cortex directs voluntary movement. It also receives and acts on messages from the **sense** organs. Functions such as **intelligence** and **memory** also involve the cerebral cortex.
The cerebral cortex is also known as grey matter.

28

cell *noun*

A cell is the basic unit of all living matter. A cell is made up of a central **nucleus** which is surrounded by **cytoplasm**. The cell is covered by cell membrane. A baby starts life as a single fertilized cell, which divides into more cells as the **embryo** develops. As the embryo grows, different kinds of cells appear, forming **skin**, **bones**, **nerves** and other parts of the body. Cells continue to divide in their different forms throughout a person's life. *An adult human body contains about 10 million million cells.*

cellular *adjective*

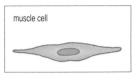
muscle cell

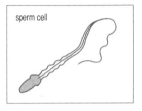
sperm cell

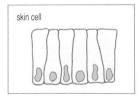
skin cell

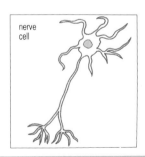
nerve cell

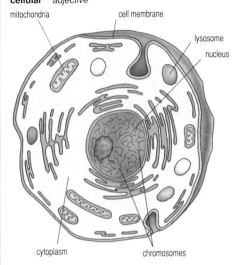

mitochondria
cell membrane
lysosome
nucleus
cytoplasm
chromosomes

A human cell

Cells in the human body are specialized according to the job they do. There are at least 100 different kinds of cell in the human body. They are all different shapes and sizes.

cerebral palsy   *noun*
Cerebral palsy is a disorder of the brain which affects posture and movement. People with cerebral palsy may not be able to speak very well, and often walk with jerky movements. Many people with cerebral palsy have normal intelligence. But they have difficulty with movement and communication. This kind of brain damage usually happens before a baby is born. It can also happen during birth or soon after a baby is born.
Cerebral palsy may be caused by a disease, an accident or illness during pregnancy.

cerebrum   *noun*
The cerebrum is the largest part of the **brain**. It is sometimes called the forebrain, as it sits in the front of the head underneath the skull. The cerebrum is made up of two half circles, or hemispheres, of soft tissue. This part of the brain controls thought and actions. It also sends messages to the **sense** organs.
Speech is controlled by three areas in the left hemisphere of the cerebrum.

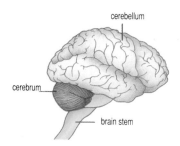

cervical vertebrae   *noun*
The cervical vertebrae are the top seven bones of the **spine**. These small, ring-shaped bones sit on top of one another to form the **neck**. The two top cervical vertebrae are called the **atlas** and **axis**. They are able to swivel around, allowing the head to turn.
The cervical vertebrae are held in place by ligaments.

cervix   *noun*
A cervix is another word for a **neck**. It can also refer to the neck of an **organ**. The cervix of the **uterus** is a narrow passageway to the **vagina**.
When a baby is ready to be born, the cervix widens to allow it to pass through.
cervical   adjective

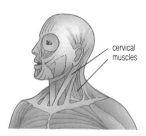

cervical muscles

Cesarean section   *noun*
A Cesarean section is an **operation** to deliver a baby. It is done under general or local **anesthetic**, so that the mother feels no pain. A doctor cuts through the wall of the mother's abdomen into the **uterus** to take out the baby. If her pelvis is too narrow for the baby to pass through, the mother must have a Cesarean section. If the baby is in the wrong position, it may be easier and safer to deliver the baby by Cesarean section.
A Cesarean section may be performed if the mother is not well.

characteristic   *noun*
1. A characteristic is another word for an essential feature of something.
A sore throat is a characteristic of a cold.
2. A characteristic is a typical feature of something else. A nose is a facial characteristic. A sense of humour is a characteristic of a personality.
When he saw the characteristic spots, the doctor knew the children had chickenpox.
characteristic   adjective

check-up *noun*
A check-up is a visit to the doctor to see that all is well. The doctor may ask a lot of questions and listen to a patient's chest and heart. Taking **blood pressure** and testing a urine sample are all part of a check-up. A doctor finds out if the patient has any health problems during this visit.
He had to have a thorough check-up before starting to train as a pilot.

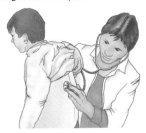

cheek *noun*
Cheeks are the fleshy parts of the face below the eyes down to the jaw. They form the outside walls of the mouth.
The skin on the baby's cheeks was smooth and soft.

chemotherapy *noun*
Chemotherapy is a treatment for a disease using a mixture of chemicals. Chemicals that are poisonous to the infection or illness are swallowed or injected into the body. Chemotherapy attacks cancer cells and destroys them.
The chemotherapy involved a mixture of several different drugs.

chest *noun*
The chest is the part of the body between the neck and the abdomen. It is also called the **thorax**. The frame of the chest is formed by 12 pairs of **ribs** which join onto the **spine**. The **heart** and **lungs** are protected by the chest.
The muscles of the diaphragm form the base of the chest.

chew *verb*
Chewing is the action of grinding food between teeth. All food must be chewed and mixed with **saliva** before it can be swallowed.
He chewed the tough meat for a long time.

chickenpox *noun*
Chickenpox is a common childhood **disease** caused by a **virus**. It is not usually a serious illness. The virus is very easy to catch and spreads quickly from one child to another. The first symptoms are a slight temperature, runny nose and sore throat. Then an itchy, red rash appears over the body. The spots soon turn into blisters. The blisters dry up into scabs after a few days. The scabs may take another week to fall off and they may leave scars which are sometimes called pockmarks.
The itching of chickenpox spots can be soothed by calamine lotion.

chilblain *noun*
A chilblain is a small, inflamed patch of skin that happens in cold, wet weather. Parts of the body which are most likely to suffer from chilblains are toes, fingers, ears and face. The skin becomes reddish-blue and swollen. It is painful, itchy and has a burning feeling. Keeping warm and dry is the best treatment for chilblains.
He wore two pairs of socks to avoid getting chilblains on his toes.

child *noun*
A child is a young human who is growing to be an **adult**. Childhood begins when a baby is learning to walk and talk. It lasts until a person starts to become sexually mature in **adolescence**. Children grow rapidly during the years of their childhood. It is a time for playing and learning all about the world around them.
The five-year-old child was learning to read at school.

childbirth ► page 32

childbirth   *noun*

Childbirth is the act of being born. Birth in humans happens about nine months after the **embryo** first begins to grow inside the mother. Birth is the moment when a baby enters the world and becomes separate from its mother.

Most babies are born head-first during childbirth.

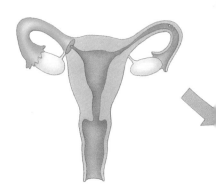

An egg, or ovum, is fertilized by a sperm. The ovum starts to develop almost immediately.

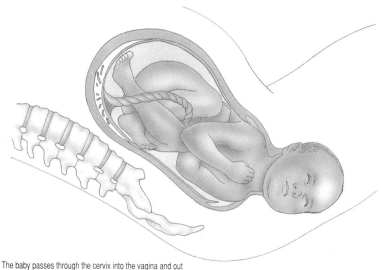

The baby passes through the cervix into the vagina and out of the mother's body.

The ovum travels down the Fallopian tube and embeds itself in the uterus.

Under eight weeks, the developing baby is called an embryo. After that time, it is called a fetus. When it is 12 weeks old, its arms, legs, hands and feet have formed, and the skeleton is developing.

The fetus lies in a bag filled with amniotic fluid. The placenta connects the bag with the uterus. The fetus obtains food and oxygen through the umbilical cord which is attached to the placenta.

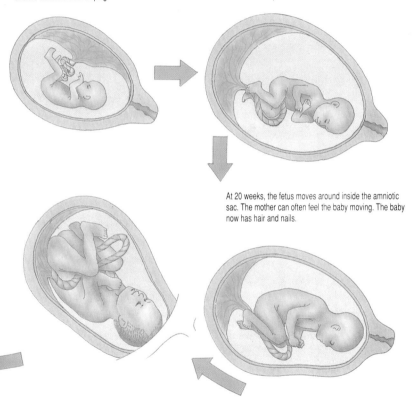

At 20 weeks, the fetus moves around inside the amniotic sac. The mother can often feel the baby moving. The baby now has hair and nails.

At 38 weeks, the baby lies upside down, ready to be born. It now fills the uterus.

chill   *noun*
A chill is the feeling of being cold. It is also an attack of **shivering**. Having a chill with shivers can be a symptom of an illness.
He caught a chill after sitting in the cold hall all afternoon.

chin   *noun*
The chin is part of the lower **jawbone**, below the mouth.
Some men have hair growing on their chin and cheeks.

chiropractic   *noun*
Chiropractic is a kind of **alternative medicine** which involves moving the bones in the **spine**. Someone who practises chiropractic is called a chiropractor. Chiropractors use their hands gently to adjust the position of the **vertebrae**.
Chiropractic can be used to relieve pain in joints and muscles.

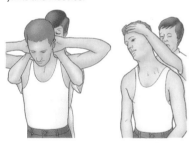

cholera   *noun*
Cholera is an **infectious disease**. It is mainly found in South America, Asia and Africa. Cholera is caused by drinking water or eating food that is infected with cholera **bacteria**. The bacteria attack the small intestine and cause severe sickness and diarrhoea. People can recover from cholera in two weeks if it is treated in time. A disease which spreads quickly like cholera is called **epidemic**.
In areas where cholera is widespread, people should boil their drinking water.

cholesterol   *noun*
Cholesterol is a fatty substance. It is used by the body to help digest food. The body makes its own cholesterol through the **liver**. It is also found in foods like butter, eggs and meat. Too much cholesterol in the **blood** can cause the **arteries** to harden. This can lead to problems with the **heart**.
Lowering cholesterol levels in the blood can help to reduce the risk of heart attacks.

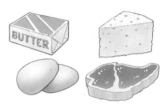

chromosome   *noun*
A chromosome is a tiny, thread-like part in the **nucleus** of a **cell**. Each chromosome carries all kinds of information about a person. Hundreds of **genes** in each chromosome store details like the colour of eyes and hair. The chromosomes in the cells of a new baby are a mixture of its parents' genes. This is why children often look like both their mother and father.
There are 23 pairs of chromosomes in each human cell.

chronic   *adjective*
Chronic describes an illness or pain which lasts a long time and changes very little. The opposite of chronic is **acute**.
Rheumatoid arthritis is a chronic disease.

circulation   *noun*
The circulation is the flow of **blood** around the body. Blood passes from the heart through **arteries**. It flows back to the heart through **veins**.
She had such bad circulation that her toes turned blue in the cold.

circulatory system *noun*

The circulatory system is the network of **veins** and **arteries** in the body. The heart, blood and blood vessels are parts of this network. Veins and arteries are strong, flexible tubes which carry blood to all parts of the body. Blood contains **nutrients** from food and **oxygen** which the body needs. It also takes away **waste products** to be flushed out of the system.

Hormones are taken round the body by the blood in the circulatory system.

circumcision *noun*

Circumcision is a small operation that is performed on the **penis**. Part or all of the **foreskin** is removed. Baby boys are often circumcised because their parents are following a family or religious tradition. Circumcision in an older child or in an adult is less common. It is usually carried out because the foreskin is too tight.

A penis works the same way, whether it has been circumcised or not.

clavicle ► collarbone

clinic *noun*

A clinic is a place that provides health advice and treatment. People visit clinics if they are ill and want to see a doctor. Patients can be treated in a clinic if they do not need to stay in hospital. Some clinics are attached to hospitals. They deal with special kinds of health problem, such as mental illness. Other clinics look after all the people in the neighbourhood.

Doctors in local clinics help people with every sort of disorder.

clitoris *noun*

The clitoris is part of the genitals of a female. It is a small piece of soft tissue just below the pubic bone. The folds of the vaginal lips, called the **labia**, protect this sensitive part of the body.

The clitoris plays an important role when a girl becomes sexually mature.

clot *noun*

A clot is a substance like jelly. It forms when a liquid thickens. **Blood** contains special substances to make it form a clot. A blood clot stops a cut from bleeding further. Some people do not have the necessary substances in their blood to make clots. These people are called hemophiliacs.

A blood clot formed at the wound, and the bleeding stopped.

clubfoot *noun*

Clubfoot is a deformity of the **foot**. Some babies are born with a clubfoot. The sole of the foot turns inwards and the heel points upwards. A clubfoot must be set in the right position with a plaster cast or metal brace.

A bad case of clubfoot can be corrected by an operation.

coccyx *noun*

The coccyx is a bone in the **spine**. It is the last bone at the bottom of the spine. The coccyx, or tailbone, is made up of four small bones joined, or fused, together.

In children, the four vertebrae of the coccyx are separate.

cochlea *noun*

The cochlea is part of the **ear**. It is also called the **inner ear**. The cochlea is a **cavity**, or space, which is shaped like a spiral and filled with fluid. **Sensory nerves** in the cochlea pick up the sounds and send messages to the **brain**.

Inside the cochlea are three tubes filled with liquid.

cochlea

cold *noun*
A cold is an infection caused by a **virus**. It is very easy to catch a cold from another person. A runny nose, headache, sore throat and cough are all **symptoms** of a cold. Ordinary colds are not serious and only last a few days. There is no cure for a cold but the symptoms can be treated with medicines.
Children catch colds more often than adults.

collagen *noun*
Collagen is a **protein**. It is the gluey kind of protein that makes up **connective tissue**, such as **tendons**. Collagen is also found in skin, bone, cartilage and ligaments.
Damage to collagen development can result in fragile, rubbery skin and loose joints.

collarbone *noun*
The collarbone, or clavicle, connects the **scapula** and **sternum**. It is shaped like an upside-down coat hanger and runs from shoulder to shoulder in the front of the body. You can feel your collarbone either side of the bottom of your neck.
When he broke his collarbone, he had to wear a sling.

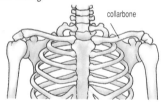

collarbone

colon *noun*
The colon is the widest part of the **intestine**. It stretches from the **cecum** to the **rectum**. The walls of the colon absorb water and salts from digested food. These are taken away by the blood and used by the body. The waste matter, or **feces**, that are left behind are stored in the rectum until they are ready to be passed through the **anus**.
The strong muscles in the colon push feces along the intestine.

colostrum ► breast-feeding

colour blindness *noun*
Colour blindness is a disorder of the **eyes**. People are born with colour blindness. If it is complete, they cannot recognize any colours at all. People with partial colour blindness cannot tell one colour from another.
If you can see a blue shape inside this circle, you are not colour blind.

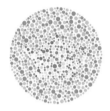

coma *noun*
A coma is a deep sleep caused by illness or injury. Comas can last days, weeks or months. Someone in a coma must be looked after in a hospital until they wake up.
Patients with diabetes can fall into a coma if their illness is not treated.

conception *noun*
Conception is a stage in sexual reproduction. It is the moment a male **sperm** fertilizes a female **egg** in a woman's **Fallopian tube**. The fertilized egg then moves into the **uterus** where it grows into a baby.
Within one day after conception, the egg begins to divide and grow.

concussion *noun*
Concussion is an injury to the **brain**, caused by a blow or an accident. Someone with concussion has usually been made unconscious. On waking up, they may feel sick and have a headache. A person with concussion should see a doctor.
He suffered mild concussion when he fell off the ladder.

cone ► eye

congenital   *adjective*
Congenital describes a disorder that a
person is born with. It can be a disease or a
deformity. Some congenital disorders are not
serious, such as **colour blindness**. Others
are more serious, such as **heart disease** or
deafness. Congenital disorders happen
when a **fetus** does not develop properly.
Down's syndrome is a congenital disorder.

conjunctiva   *noun*
The conjunctiva is a thin **mucous
membrane** that covers the front of the **eye**.
It also lines the inside of the **eyelid**.
*The conjunctiva produces a fluid that
moistens and protects the eyeball.*

conjunctivitis   *noun*
Conjunctivitis is an inflammation of the
conjunctiva of the eye. Someone with a cold
or influenza may develop conjunctivitis. The
white of the eye becomes bloodshot and
sore. The eyes water and are sensitive to
bright light. Conjunctivitis may be caused by
an **allergy**, or by **bacteria** or **viruses**.
*Conjunctivitis caused by bacteria can be
treated with antibiotic eye drops.*

connective tissue   *noun*
Connective tissue joins and holds parts of
the body together. **Tendons** and **ligaments**
are different kinds of connective tissue.
They connect and support **bones** and
muscles.
*Connective tissue binds together bunches of
muscle fibre.*

constipation   *noun*
Constipation is a bowel condition. A person
has constipation when they find it difficult to
pass out **feces** through the **anus**. How often
the body needs to pass feces varies from
person to person. Some people have bowel
motions every day, others only three times
a week.
Eating fibre helps to prevent constipation.

contact lens   *noun*
A contact lens is a round piece of glass or
plastic that fits over the **cornea** of the **eye**.
Contact lenses are worn to correct a sight
defect, when the person cannot see as well
as they should. People wear contact lenses
instead of spectacles. Unlike spectacles,
contact lenses move as the eye moves.
There are two kinds of contact lens — hard
and soft.
He kept his soft contact lenses in all day.

contagious   *adjective*
Contagious describes a **disease** which is
caught from another person by touching that
person. A contagious disease can also be
caught by contact with cups, plates or other
objects used by an infected person. Animals
and insects can also spread contagious
diseases.
*Chickenpox, measles and influenza are all
contagious diseases.*
contagion   *noun*

contraception   *noun*
Contraception is a way of stopping a **sperm**
fertilizing an **ovum** to make a baby. It is used
by men and women who want to choose
whether they have children or how many.
There are many different ways of preventing
conception. Fitting a special cap over the
cervix stops sperm entering the **uterus**.
A rubber sheath, or condom, can be fitted to
catch all the sperm.
*Hormone pills are an effective method of
contraception.*
contraceptive   *noun and adjective*

convulsion *noun*
A convulsion is a fit of muscle **spasms**. Someone having a convulsion is usually **unconscious**. Their body jerks violently as the muscles contract and relax. A high **fever** can cause convulsions, especially in children under two years old. **Epilepsy** causes regular convulsions. Convulsions are not harmful in themselves and a person recovers quickly.
They moved the furniture to stop her hurting herself during the epileptic convulsion.

co-ordination *noun*
Co-ordination is being able to time movements in the right order. Walking needs co-ordination. Every part of each leg has to move to the right place at the right time. The **cerebellum** controls the body's co-ordination.
The ballet dancer needed superb co-ordination to perform dramatic leaps.

corn *noun*
A corn is a small area of thick skin on the foot. Hard corns are painful when the core inside presses on a nerve. Corns between the toes are soft and sometimes become inflamed. Most corns are caused by wearing shoes that do not fit properly. Hard corns can be removed by a chiropodist, a person who looks after people's feet.
Corn plasters are used to relieve pain from corns.

cornea *noun*
The cornea is part of the **eyeball**. It is the clear layer of **membrane** which covers the front of the eye, underneath the **conjunctiva**. Light rays focus through the cornea to the **lens** of the eye.
There are no blood vessels in the cornea.

coronary *adjective*
Coronary is a word that describes anything to do with the heart.
Coronary arteries supply blood to the heart.

coronary heart disease ► heart disease

coronary thrombosis *noun*
A coronary thrombosis is a heart attack. It happens when the blood supply to the heart is partly or completely cut off. This takes place when the coronary **arteries** become blocked, usually by a blood **clot**. People can recover from a mild coronary thrombosis. More serious ones are fatal.
Regular exercise and a healthy diet can reduce the risk of coronary thrombosis.

corpuscle *noun*
A corpuscle is a **cell** or other small, rounded body.
Red and white blood cells are corpuscles.

cortex *noun*
A cortex is the outer layer of an **organ**. The outer layer of the brain is called the **cerebral cortex**.
The adrenal cortex secretes hormones.

cortisone *noun*
Cortisone is a substance made by the body. It is a **hormone** that is made in the outer layer, or **cortex**, of the **adrenal gland**. The liver turns cortisone into cortisol, which helps to control glucose, fats and water in the body. Doctors use an **artificial** cortisone to treat a number of diseases.
She had a course of cortisone injections to relieve her arthritis.

cough *noun*
A cough is a sudden push of air out of the **lungs**. It is the body's way of clearing the lungs or **throat**. A cough may be a **symptom** of a **disease** or **illness**, such as a chest **infection** or **asthma**.
The smoky atmosphere gave him a cough.

cramp *noun*
A cramp is a sudden tightening, or **spasm**, in a **muscle** which causes pain. Too much exercise can cause cramp. Sweating and losing a lot of body fluids is another reason for cramp. Gently stretching and warming the muscle helps to ease a cramp.
The athlete had cramp in his leg after the race.

cranium ► skull

croup *noun*
Croup is a breathing disorder. It is most common in children aged between one and four years. Virus infections are the main cause of croup. The **lungs**, **throat** and **voice box** become inflamed and swollen. Children with croup have a hoarse cough which sounds like a bark. They have a **fever** and cannot breathe easily. If a croup attack is mild, a child can be treated at home. A steamy atmosphere will ease the breathing.
His attack of croup lasted three days.

crown *noun*
A crown is the part of a tooth you can see. A dentist can fit an **artificial** crown on to a damaged tooth. Some of these artificial crowns are made of gold.
The artificial crown that the dentist made blended in perfectly with the boy's other teeth.

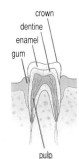
crown
dentine
enamel
gum
pulp

cut *noun*
A cut is an opening in the skin made by a sharp object. When the skin is cut, the body bleeds. Pressing firmly on a cut stops the bleeding. If **germs** enter the body, a cut may become **infected**. When a cut is long and deep it may need to be stitched up by a doctor. This helps it to heal. A surgeon makes a cut in the skin during an **operation** using a special knife called a scalpel.
Most cuts heal very quickly by themselves.

cuticle *noun*
The cuticle is an outer layer of **skin**. It is also the tough skin found at the base of **fingernails** or **toenails**.
She pushed down the cuticles of her fingernails.

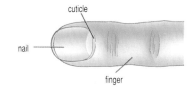
cuticle
nail
finger

cyst *noun*
A cyst is a swelling or lump filled with fluid. Cysts commonly form in the **skin**, **breasts** and **ovaries**. Cysts can develop if a gland becomes blocked. Cysts are **benign** and do not lead to serious illness. In some cases, a doctor drains the fluid from a cyst with a needle and the swelling disappears.
If a cyst is large or painful, it can be removed by an operation.

deafness *noun*
Deafness is not being able to hear well. It can affect one or both ears. People can be born deaf, or deafness can develop in later life. Elderly people often develop some deafness. Partial deafness can be helped by a hearing aid.
The deaf boy learned to communicate by using sign language.

death *noun*
Death is the end of life. Everyone dies eventually. When people live for a long time, parts of their body stop working and they die. This is part of the **aging** process. Death can also be caused by an illness or an accident.
After the death of his father, he took over the family business.

dehydration *noun*
Dehydration is the result of losing too much water from the body. Severe **diarrhea**, **vomiting** and **sweating** cause dehydration. Very hot weather or having a high fever may make someone sweat too much.
People suffering from dehydration should drink plenty of liquid.

dementia *noun*
Dementia is a brain illness. A person with dementia cannot remember things or think clearly. Eventually, they may have a change of personality. Elderly people sometimes develop dementia, or it can be caused by a **tumour** in the brain.
The old lady's dementia made her confuse night-time with daytime.

dental *adjective*
Dental is a word that describes anything to do with **teeth**. Dental hospitals look after people's teeth and gums.
The dental nurse assisted the dentist with the fillings.

dentine *noun*
Dentine is the main part of a **tooth**. It is the hard substance that surrounds the soft pulp at the centre of the tooth. Dentine is covered by a layer of tough **enamel**. Enamel is the white, outer part of the tooth that can be seen.
Dentine is harder than bone.

dentist *noun*
A dentist is someone who looks after people's **teeth**, **mouth** and **gums**. Dentists drill away decay in teeth and fill in the holes. They take out badly decayed teeth which are painful and can fit **artificial**, or false, teeth. Dentists tell people how to avoid tooth decay by eating a good diet and cleaning their teeth properly.
Dentists advise people to have a dental check-up at least once a year.

dentition *noun*
Dentition is the development of **teeth**. Babies' teeth appear during their first year. These first teeth are called milk teeth. There are 20 milk teeth. These start to become loose and fall out one by one from about the age of six years. They are gradually replaced with a set of 32 adult teeth and dentition is complete.
Dentition made the baby dribble.

depression   *noun*
Depression is a mental disorder. People
suffering from depression may feel sad or
helpless. They often lose interest in what is
going on around them. Depression can be
treated by psychotherapy or **drugs**.
*While she was suffering from depression,
she kept bursting into tears.*

dermatitis   *noun*
Dermatitis is **inflammation** of the **skin**. The
symptoms are dry, itchy patches of red skin
which can become very sore if the skin
breaks and becomes infected. Dermatitis
may develop when the skin has been in
contact with a chemical substance, such as
detergent or acid. Some metals and a few
plants also cause dermatitis.
Dermatitis can be treated with special creams.

dermis   *noun*
The dermis is the middle layer of the skin.
It lies below the top layer of skin, or
epidermis. The dermis contains connective
tissue, blood vessels, nerve endings and hair
follicles.
The dermis is thicker than the epidermis.

develop   *verb*
Develop is another word for grow. Children
develop into adults. To develop also means
to change and become something else.
A cold may develop into pneumonia.
She developed a rash after lying in the sun.

diabetes   *noun*
Diabetes is the name for two separate
disorders. Diabetes insipidus is caused by
the lack of a special **hormone** called ADH.
Diabetes mellitus is caused by the lack of an
essential hormone called **insulin**. People
who suffer from diabetes are called diabetics.
The **symptoms** of diabetes are passing a lot
of urine and feeling thirsty all the time.
Without treatment a diabetic will become
very ill and fall into a coma.
*She developed diabetes when she was eight
years old.*

diagnosis   *noun*
A diagnosis is the decision a doctor makes
about what kind of disorder his patient is
suffering from. To make a diagnosis, the
doctor looks for **symptoms** and signs of
illness. Patients are examined and asked
questions about how they feel. They may
have medical tests, such as an **X-ray** or a
blood test.
*Doctors make a diagnosis so that they can
give a patient the right treatment.*
diagnose   *verb*

dialysis   *noun*
Dialysis describes a method of cleaning the
blood with a machine. This job is usually
done by the **kidneys**. A dialysis machine
works like a kidney. Blood is let out of an
artery into the machine. It passes through a
filter which takes out unwanted substances.
The clean blood is then fed back into a **vein**.
Dialysis usually needs to be carried out three
times a week.
*Cleaning all a patient's blood by dialysis
takes between four and six hours.*

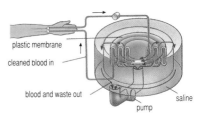

plastic membrane

cleaned blood in

blood and waste out

saline

pump

diaphragm   *noun*
The diaphragm is a large sheet of **muscle**.
It separates the organs in the **abdomen** from
the heart and lungs in the chest. The
diaphragm is the most important muscle
used for breathing. It tightens and becomes
flat to draw breath into the lungs. Air is
pushed out when the diaphragm relaxes and
expands upwards in a curve.
*The diaphragm muscle is shaped like a
dome.*

digestion *noun*

Digestion is what happens to **food** in the body. Food starts to be digested as soon as it enters the mouth. It is broken down into smaller and smaller pieces as it travels through the digestive system. The digestive organs include the **stomach**, the **small intestine** and the **large intestine**.

Vitamins and minerals do not need digestion as they are absorbed straight into the bloodstream.

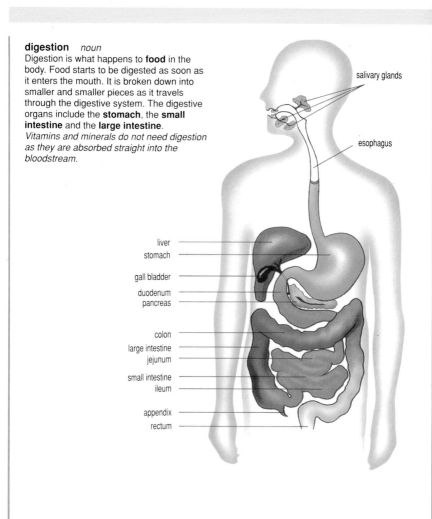

salivary glands

esophagus

liver
stomach

gall bladder
duodenum
pancreas

colon
large intestine
jejunum

small intestine
ileum

appendix
rectum

1. As food is ground by the teeth, it is mixed with saliva. Saliva is produced by three glands in the face and contains enzymes. Each mouthful of food is rolled into a ball that is easy to swallow.

2. The ball of food travels down the esophagus to the stomach. The elastic walls of the stomach expand as it fills with food. Gastric juices from glands in the stomach wall mix with the food and start to digest it. The juices contain hydrochloric acid. The stomach churns to mix the contents thoroughly. Food stays in the stomach between two and five hours.

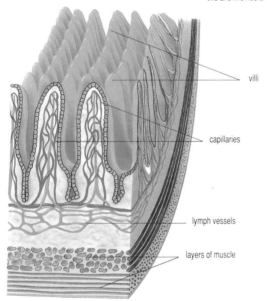

villi

capillaries

lymph vessels

layers of muscle

3. The partly-digested food is now a thick liquid called chyme. It enters the duodenum where it is mixed with juices from the pancreas, liver and gall bladder.

4. In the jejunum and the ileum, nutrients from digested food are absorbed into the bloodstream.

5. Digestion is almost complete by the time food reaches the large intestine. The undigested food is pushed down the colon. Water and some minerals are absorbed in the colon.

6. Bacteria feed on the waste and turn it into feces, which then pass down the rectum, and out through the anus.

The walls of the small intestine are lined with tiny fingers called villi. The mass of villi make the surface look like velvet. Villi contain blood capillaries and other vessels that contain lymph. The lymph vessels absorb fat and the capillaries take nutrients from the digested food into the bloodstream.

diarrhea *noun*

Diarrhea is soft, liquid **feces** that are passed frequently. It can be the result of a mild stomach upset or it may be a **symptom** of a more serious disease. Diarrhea makes the body lose a lot of water which can cause **dehydration**. People with diarrhea should drink plenty of fluids.

He had diarrhea after eating the shellfish.

diet *noun*

1. A diet is the mixture of foods a person eats every day. To keep the body healthy a diet must include **proteins**, **carbohydrates**, **fats**, **vitamins**, **fibre** and **minerals**.

A varied diet gives the body what it needs.

2. A diet is a way of controlling the intake of food. People may go on a diet because they want to lose weight. Some people who suffer from certain diseases follow a special diet.

After his heart-attack, the doctor recommended a low fat diet.

digestion ► page 42

diphtheria *noun*

Diphtheria is an **infectious disease** that mainly affects breathing. It is more common in children under the age of 10. Diphtheria is caused by **bacteria** which grow in the nose and throat. Diphtheria is now a rare disease because of **vaccination**. The **vaccine** is given to babies in their first year to protect them.

Doctors recommend three doses of diphtheria vaccine in the first year.

disability *noun*

A disability describes the loss of function of part of the body. Some people are born with a disability, and some acquire it through accident or illness. The disability may be mental or physical. Special equipment and other benefits help disabled people to live as normal a life as possible.

Her disability did not stop her joining the team.

disabled *adjective*

disc *noun*

A disc is a round, flat, plate-shaped object. In the body, the optic disc is found at the back of the **eye**. Another kind of disc lies between each pair of **vertebrae** in the **spine**. These act as a cushion between the bones and allow the spine to bend easily.

Spinal discs are made of cartilage with a soft, jelly-like centre.

disease ► page 45

disinfectant *noun*

Disinfectant is a chemical which is used to kill **germs**. Destroying germs stops the spread of infection. People put disinfectant in water to wash areas such as floors and toilets where there are a lot of germs.

She washed the thermometer in disinfectant after taking his temperature.

dislocation *noun*

Dislocation describes what happens when a joint is knocked out of place. This can happen through an accident or injury and causes great pain. A doctor should always be called to put the joint back in place.

Finger joints are the most common joints to suffer dislocation.

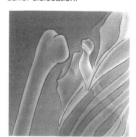

An X-ray of a dislocated shoulder

dissect *verb*

Dissect is to cut up carefully. Students dissect parts of plants and animals to learn about their **anatomy**.

She dissected the dead mouse to examine its lungs.

disease   *noun*

Disease is a sickness in some part of the body or the mind. Many people have mild diseases such as colds, and recover from them. Other diseases can be incurable, or fatal. Sometimes people have a disease that they have inherited. Some diseases are **infectious**. Other diseases happen when the body or mind is damaged in some way. *The builder developed lung disease from working with asbestos.*

Muscular dystrophy is an inherited disease. It causes gradual weakening of the muscles until a person can no longer move around on their own.

tetanus bacteria

Some diseases, such as tetanus, develop when bacteria enter the body through a cut in the skin. Tetanus bacteria may be found in the soil.

leukemia cells

Leukemia is a disease of the blood. It is a type of cancer and is not infectious.

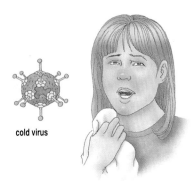

cold virus

Infectious diseases are spread by people, by animals and insects, and other objects such as food. Colds are easily spread to other people.

Diseases are treated in many different ways. The body's immune system fights disease, or a doctor may prescribe drugs or surgery. Many diseases can be prevented by good hygiene, a healthy lifestyle, and immunization.

dizziness *noun*
Dizziness is a feeling of being light-headed and unsteady. Dizziness may happen when a person's sense of balance is upset, or when not enough blood reaches the brain.
Repeated attacks of dizziness may be a **symptom** of an illness.
Feeling hungry or tired sometimes causes dizziness.

DNA *noun*
DNA is a chemical found in the middle, or **nucleus**, of every **cell**. The letters DNA stand for deoxyribonucleic acid. DNA stores and sends on **genetic** information to new cells. The genetic information gives every detail of how a person is put together and what they look like. DNA provides the master plan, or **blueprint**, of each person.
Everyone except identical twins has a different kind of DNA.

doctor *noun*
A doctor is a person who looks after people when they are ill and helps them to keep healthy. Men and women have special medical training to become doctors. Family doctors deal with all kinds of health problems. Other doctors, or specialists, treat only certain kinds of illness, or groups of people. A specialist doctor who looks after children is called a paediatrician.
The doctor advised her to stay in bed for four days.

dorsal *adjective*
Dorsal describes the back, or **posterior**, of any part of the body. The opposite of dorsal is **ventral**.
Dorsal roots lie at the back of the spinal cord.

dose *noun*
A dose is a measure of **medicine**.
All medicines are carefully mixed and measured. If a person is taking medicine they must have the right dose.
He had to take a dose of medicine three times a day.

Down's syndrome *noun*
Down's syndrome is a disorder that people are born with. People with Down's syndrome have an extra **chromosome** in their cells. This makes them physically and mentally less able than other people. People with Down's syndrome take much longer to learn how to do things for themselves. They can often be helped by careful training and support, and lead enjoyable lives.
People with Down's syndrome often have very loving and happy personalities.

dream ► **sleep**

drug *noun*
A drug is a chemical substance. Organic drugs are found in plants or other living material. Synthetic drugs are made from **artificial** substances. Drugs are used to make **medicines**. Different kinds of medicine relieve or cure all kinds of illness. They are also used to prevent illness. Drugs are harmful if they are not used properly. They should only be taken on a doctor's advice.
The scientists kept the drugs locked up safely.

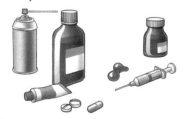

duct *noun*
A duct is a narrow tube in the body. Ducts carry fluids from one part of the body to another.
Tears flow into the eyes through the tear ducts.

duodenum *noun*
The duodenum is part of the **small intestine**. The small intestine is about six metres long, and the duodenum is the first part. Partly-digested food passes from the **stomach** to the duodenum. Digestive juices from the **pancreas** and **gall bladder** help to break down the food in the duodenum.
The duodenum absorbs nutrients and water.

dysentery *noun*
Dysentery is an **inflammation** of the **large intestine**. The inflammation causes severe **diarrhea** which may contain blood and mucus. There may also be pains in the abdomen, sickness and fever. Dysentery is caused by **bacteria** or a **parasite**. These can be picked up by drinking contaminated water or eating infected food.
Improvements in hygiene have made dysentery much less common than it used to be.

dyslexia *noun*
Dyslexia is a specific learning disorder. It affects the way a child learns how to read and write, but not their **intelligence**. Children with dyslexia muddle up their letters and find spelling difficult. Special teaching helps the condition.
Many people with dyslexia cannot tell left from right.

ear ► page 48

earache *noun*
Earache is a pain inside the **ear**. It can be caused by an **infection** in the middle ear or **inflammation** in the outer ear. Sudden changes in air pressure, such as in an aeroplane, can cause earache. Earache may also be caused by the ear being blocked with wax, or other objects.
He had earache when he went diving.

eardrum *noun*
The eardrum is part of the **ear.** It is a thin layer of **tissue**, called the tympanic membrane, that separates the middle ear from the outer ear.
Very loud noises can damage the eardrum and cause deafness.

eat *verb*
To eat is to take food into the body through the mouth. People need to eat a balanced **diet** of food to stay healthy. Not eating properly can lead to an eating disorder, such as **anorexia** or **bulimia**.
A vegetarian does not eat meat.

eczema *noun*
Eczema is **inflammation** of the **skin**. It can appear anywhere on the body in children or adults. The skin develops red patches and feels itchy and sore. Most people suffer from eczema because it runs in the family, or is inherited. Eczema may develop from contact with chemicals or an allergy.
Whenever he was under stress, his eczema became worse.

ear *noun*

The ear is the organ of **hearing** and **balance**. People have two ears, one on each side of the head. The **outer ear** picks up sounds and passes them to the **middle ear** through the **eardrum.** Three tiny bones in the middle ear vibrate with the sound. These vibrations are passed to the **cochlea** in the **inner ear**, where they are changed into electric signals. The signals travel along **nerves** to the **brain**.

Hearing with two ears helps people to tell what direction a sound is coming from.

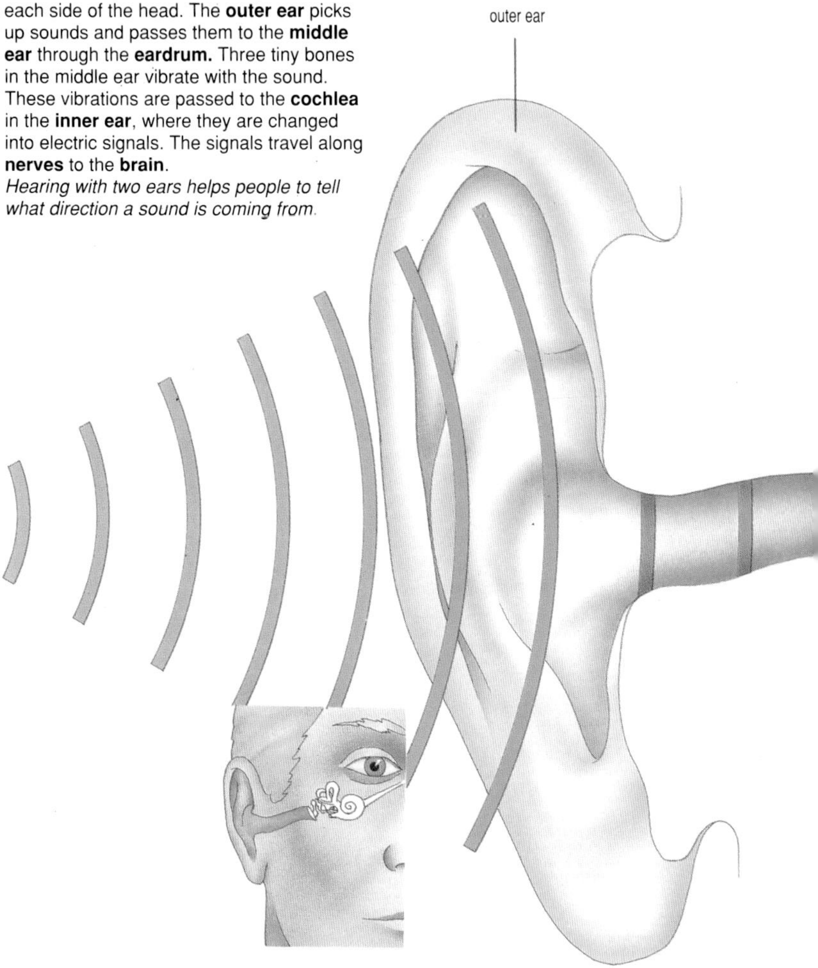

outer ear

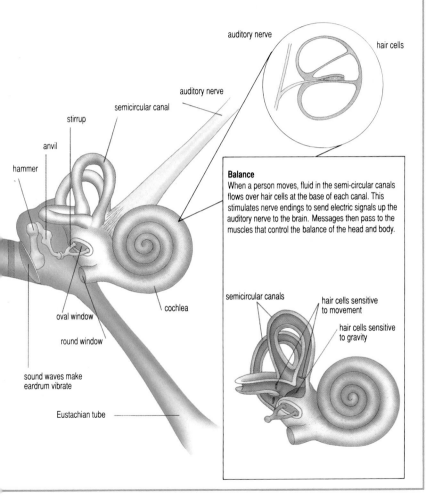

Sound vibrations move the fluid in the cochlea. This stimulates hair cells in the central section of the cochlea. Messages travel to the brain up the auditory nerve.

auditory nerve

hair cells

auditory nerve

semicircular canal

stirrup

anvil

hammer

Balance
When a person moves, fluid in the semi-circular canals flows over hair cells at the base of each canal. This stimulates nerve endings to send electric signals up the auditory nerve to the brain. Messages then pass to the muscles that control the balance of the head and body.

semicircular canals

hair cells sensitive to movement

hair cells sensitive to gravity

oval window

cochlea

round window

sound waves make eardrum vibrate

Eustachian tube

egg ▶ ovum

elbow   *noun*
The elbow is part of the **arm**. It is the **joint**
that joins the upper and lower bones of the
arm. This joint works like a hinge and allows
the arm to bend in half. Two **muscles** in the
upper arm, called the **biceps** and **triceps**,
help the arm to bend at the elbow joint.
He grazed his elbow when he fell over.

electrocardiograph   *noun*
An electrocardiograph is a machine. It
records electrical activity from the **heart**.
Small metal plates called electrodes are
placed on the body. The electrodes pick up
electric signals. The pattern of the signals is
then scribbled on to a moving strip of paper.
A doctor can see if a person's heart is
working properly by looking at the pattern.
*The electrocardiograph showed that her
heart was beating normally.*

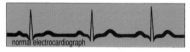
normal electrocardiograph

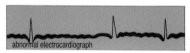
abnormal electrocardiograph

electroencephalograph   *noun*
An electroencephalograph is a machine that
records electrical activity from the **brain**. This
machine works in the same way as an
electrocardiograph.
*The electroencephalograph showed that he
had a tendency to have fits.*

elephantiasis   *noun*
Elephantiasis is a **skin** disease. It is caused
when a person is bitten by a mosquito that is
infected by a tiny **worm**. Rough skin
develops, with **swelling** in the body, usually
the legs.
*Elephantiasis is most common in tropical
countries.*

embolism   *noun*
Embolism describes when a blood vessel,
such as an **artery**, becomes blocked. The
blockage is called an embolus. An embolus
can be a bubble of air or piece of body
tissue, but it is most often a blood **clot**. The
clot is carried around the bloodstream until it
sticks somewhere. If a blood clot becomes
lodged in the blood vessels in the head, it
can result in a **stroke**.
*Embolism can be treated with drugs or
surgery.*

embryo   *noun*
An embryo is a tiny, unborn baby. Embryo is
the medical name for a fertilized **ovum** that
has settled into the **uterus** to grow. After the
second month of **pregnancy** the embryo is
called a **fetus**.
*An embryo has tiny stumps which become
the arms and legs.*

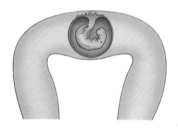

embryology   *noun*
Embryology is the study of unborn babies.
*Embryology explains how a baby grows from
egg to fetus.*

emphysema   *noun*
Emphysema is a disease of the **lungs**. The
air sacs of the lungs, or **alveoli**, do not
breathe out all the air in them. The lungs
become over-stretched with trapped air. A
person with emphysema finds it difficult to
breathe and feels tight around the chest.
Smoking cigarettes causes emphysema.

enamel   *noun*
Enamel is the outer layer of a **tooth**. It is a hard, white substance that covers the part of the tooth that can be seen, or the **crown**. **Fluoride** can help to strengthen tooth enamel.
She chipped the enamel on her tooth when biting on the hard sweet.

encephalitis   *noun*
Encephalitis is an inflammation of the **brain**. It is usually caused by a **virus** or **bacteria**. The first symptoms are very severe headaches, a fever and sickness. A person with encephalitis may feel drowsy and confused. If the illness is severe they may fall into a coma and have convulsions. **Antibiotic drugs** help to cure the disease if it is caused by bacteria.
One kind of encephalitis is known as sleeping sickness.

endemic   *adjective*
Endemic describes a **disease** that is only found in one part of the world. It also describes types of illnesses only found in a certain group or race of people.
Malaria is a disease which is endemic in tropical countries.

endo-   *prefix*
Endo- before a word means inside or mixed in with. An endoscope is an instrument for looking inside the body. The opposite of endo- is ecto-, which means outside.
The endocardium is a membrane which lines the inside of the heart.

endocrine gland   *noun*
An endocrine gland is an organ that makes **hormones**. Some endocrine glands store hormones. These kinds of gland do not have tubes, or **ducts**. They release hormones straight into the bloodstream. These hormones help to control other organs in the body.
Endocrine glands produce hormones that make the body work properly.

endocrine system   *noun*
The endocrine system is a set of glands in the body. There are seven kinds of endocrine glands inside the head, neck, abdomen and genitals. The endocrine glands release **hormones** which make the body work properly. Many important functions of the body, such as reproduction, are organized by hormones. The **pituitary gland** at the base of the brain controls the release of hormones.
Growth and digesting food are part of the work of the endocrine system.

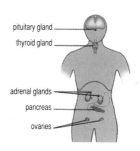

pituitary gland
thyroid gland
adrenal glands
pancreas
ovaries

endorphin   *noun*
Endorphin is a **hormone**. It is made in the brain. Endorphins are released in the body to help deaden pain. They also make a person feel happy. Exercise makes the body release endorphins to relieve aching muscles.
Endorphins help the body to cope with stress.

endoscope   *noun*
An endoscope is a medical instrument. It is a long, bendy tube with a light on the end and is used to look inside the body. The endoscope can be pushed into a natural opening of the body, such as the nose or mouth. The stomach and lungs are often looked at in this way.
The doctor used an endoscope to examine the patient's stomach.

energy *noun*
Energy is power and strength. People burn up energy like fuel as they live. **Food** has an energy value which is measured in **calories**, or kilojoules. Different foods have different calorie, or energy values. An adult man needs about 2,500 calories of energy a day to live. People who are extra active, such as athletes, need more calories. People become overweight if they take in more calories than they burn up.
Extra energy is stored in the body as fat.

environment *noun*
Environment is a word that describes a person's surroundings. It may be a place, such as a town. Or it may be living conditions, such as those found at home, school and work. Having a clean environment is important to health. People need clean air, water and food. They also need warmth, shelter and space.
Pollution threatens the environment of people who live on our planet.

enzyme *noun*
An enzyme is a **chemical** made by the body. It joins other chemicals to help them work faster. For example, enzymes are part of digestive juices in the **stomach** which help break down food.
The human body contains over a thousand types of enzyme.

epi- *prefix*
Epi- before a word means over, above or upon. The epidermis covers the dermis, or middle layer of skin.
The epithelium is a thin layer of tissue that covers surfaces inside the body.

epidemic *noun*
An epidemic is an outbreak of **infectious disease**. Epidemics spread quickly between people in the same place at the same time. **Cholera** is an epidemic disease.
An epidemic of influenza closed the school.
epidemic *adjective*

epidermis *noun*
The epidermis is the outer layer of the **skin**. It is a waterproof layer of **tissue** that grows and sheds cells all the time. Sweat passes through tiny openings, or **pores**, in the epidermis. Hair appears through the epidermis all over the body, apart from the palms of the hand and the soles of the feet. The epidermis covers and protects the middle layer of skin, or **dermis**.
Most household dust is made of old skin cells shed from the epidermis.

epididymis *noun*
The epididymis is a structure on the outside of the **testis**. It is a very thin tube about six metres long, coiled very tightly inside the **scrotum**. The epididymis is joined to the testis by small tubes.
Sperm grows in the epididymis.

epidural *adjective*
Epidural describes anything to do with the outer **membrane** that surrounds the **brain** and the **spinal cord**.
She was given an epidural anesthetic before the operation.

epiglottis *noun*
The epiglottis is part of the throat. It is a thin, leaf-shaped flap which lies behind the root of the **tongue**. The epiglottis covers the opening of the **larynx** and the **trachea** during swallowing. It stops food and drink going down the wrong hole into the lungs.
Choking can happen when a piece of food gets past the epiglottis into the trachea.

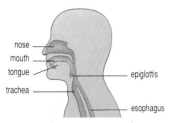

nose
mouth
tongue
trachea
epiglottis
esophagus

epilepsy   *noun*

Epilepsy is a disorder of the **brain**. People who suffer from epilepsy have regular fits, or **convulsions**. Epilepsy may affect part or all of the brain. Partial epilepsy shows as a slight twitching in the face, arm or leg. The person stays awake but seems far away. Generalized epilepsy leads to convulsions and loss of consciousness. Epilepsy is caused by a surge of electrical activity in the brain.
He was not able to drive a car because he suffered from epilepsy.

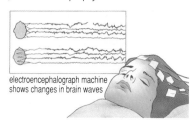
electroencephalograph machine shows changes in brain waves

equilibrium ► balance

erythrocyte ► red blood cell

esophagus   *noun*

The esophagus joins the back of the **pharynx** to the **stomach**. It is a tube that measures about 25 centimetres long. Food and drink are passed down the esophagus by muscles moving in waves. Another name for the esophagus is the gullet.
Heartburn is caused by inflammation of the esophagus.

estrogen   *noun*

Estrogen is a female **hormone**. It is made by the **ovaries** and the **adrenal glands**. Estrogen helps a girl develop into a woman. It causes the breasts to develop and causes other changes at **puberty**. Estrogen controls the monthly cycle of the uterus, or **menstruation**.
Artificial estrogen is one of the ingredients of the contraceptive pill.

Eustachian tube   *noun*

The Eustachian tube joins the back of the nose with the **middle ear**. It is about four centimetres long. The Eustachian tube keeps the air pressure the same on each side of the **eardrum**. A blockage of the Eustachian tube can lead to **deafness**.
The Eustachian tube is made of bone and cartilage.

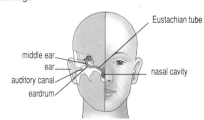

middle ear
ear
auditory canal
eardrum
Eustachian tube
nasal cavity

excretion   *noun*

Excretion is the removal of **waste** from the body. Getting rid of **urine** and **feces** is part of the process of excretion. Sweating through the skin and breathing out carbon dioxide from the lungs is also excretion.
One third of a teaspoonful of salt is removed from the body by excretion every day.
excrete   verb

exercise   *noun*

Exercise is all kinds of physical activity. It keeps the body in good working order. Exercise makes people strong and supple, and also builds up stamina. Stamina is being able to do things for a long time and not get tired. People need to exercise at least three times a week for half an hour to stay healthy. Being fit helps fight off illness. Taking regular exercise also makes people happier, more alert and more relaxed.
Swimming is the best exercise for strength, stamina and suppleness.
exercise   verb

exhale ► respiration

expire ► respiration

53

eye *noun*

The eye is the organ of sight. People have two eyes in front of the face. The objects that people see give off, or reflect, light. Light enters the eye through the **pupil**, and travels to the **retina**. There it is converted into nerve signals that travel to the brain.
Having two eyes helps people to judge distances.

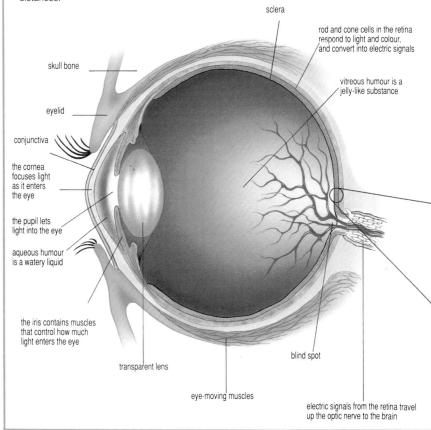

sclera

rod and cone cells in the retina respond to light and colour, and convert into electric signals

vitreous humour is a jelly-like substance

skull bone

eyelid

conjunctiva

the cornea focuses light as it enters the eye

the pupil lets light into the eye

aqueous humour is a watery liquid

the iris contains muscles that control how much light enters the eye

transparent lens

blind spot

eye-moving muscles

electric signals from the retina travel up the optic nerve to the brain

Light rays from a distant object travel to the eye in straight lines. The cornea and the lens bend the light rays to make them come together on the retina. The lens is flat and relaxed.

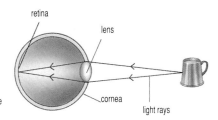

Light rays from a nearby object spread out as they enter the eye. To focus the rays, the lens must bend them more sharply. The lens becomes rounder and thicker.

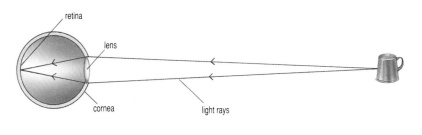

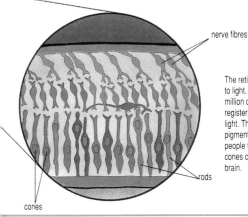

The retina is made up of a mass of cells that are sensitive to light. There are about 120 million rod cells and about 6 million cone cells. Rods are sensitive to light, but only register shades of grey. They help people to see in dim light. The cones only work well in good light. They contain pigments that absorb green, red and blue light, and enable people to see colour. Nerve fibres attached to the rods and cones carry electric signals to the optic nerve and to the brain.

external   *adjective*
External describes anything outside, or the outside part of something.
The skin is an external organ of the body.

eye ▶ page 54

eyeball·   *noun*
The eyeball is the globe of the **eye**. It is almost a perfect sphere, about 2.5 centimetres across and filled with a jelly-like substance, called **vitreous humour**. The eyeballs lie in the two spaces, or **orbits**, in the front of the skull. Three sets of **muscles** hold each eyeball in place and allow it to move around. Eyeballs are covered by a thin layer of tissue called the sclera.
Tears washed the dust off her eyeball.

eyebrow   *noun*
Eyebrows are the hairy arches that curve over each **eye**. They follow the bony ridges of the top of the **orbits** in the skull where the eyeballs lie. Eyebrows help to protect the eye from dust. They also shade the eye from light.
He raised his eyebrow in surprise.

eyelash ▶ **eyelid**

eyelid   *noun*
An eyelid is part of an **eye**. Both eyes in the body have top and bottom eyelids fringed with hair, or eyelashes. Eyelids are coverings that move up and down to protect eyes. They are made of **skin**, **muscle** and **connective tissue**. Eyelids blink, or close quickly every few seconds to keep the eye moist with **tears**. They also close the eyes during sleep.
A person blinks the eyelids once every two to ten seconds.

face   *noun*
The face is the front, or **anterior**, part of the **head**. Parts, or features, of the face include the forehead, a pair of eyes, eyebrows and cheeks, a nose, mouth and chin.
She looked at her face in the mirror.
facial   *adjective*

faint   *verb*
Faint describes becoming unconscious for a short time. A person may suddenly feel dizzy, sick and weak and the next moment fall to the ground. Fainting is caused by the **brain** not getting enough blood. Falling over makes a fainting person wake up quickly as the blood rushes back to the head. Fainting is a common **symptom** and not usually serious in itself.
She had eaten so little food, she felt she was about to faint.
faint   *noun*

Fallopian tube   *noun*
The Fallopian tubes are two tubes in a woman's body. Each tube is about 10 centimetres long and attached to either side of the top of the **uterus**. Eggs travel from the **ovaries** down the Fallopian tubes, helped along by tiny, hair-like **cells**. If a male **sperm** meets an egg in one of the Fallopian tubes, **fertilization** takes place. The fertilized egg, or **embryo**, carries on down the Fallopian tube until it settles into the lining of the uterus and starts to grow.
Conception usually happens in the Fallopian tubes.

far-sightedness ▶ **long-sightedness**

fat *noun*
Fat is a kind of food. It gives the body **energy**, together with **protein** and **carbohydrate**. Fat is found in butter, cream, margarine and oil. It is also found in fatty meats and oily fish. Fats that are not burned up and used as energy are stored in the body as fat **cells**. Some stored fats surround and protect **organs** inside the body.
A healthy diet contains between 15 and 25 grams of fat a day.

fat cells

fatigue *noun*
Fatigue is the feeling of being exhausted, or very tired. It can be tiredness of the body or mind. Fatigue may be caused by heavy exercise or lack of sleep. Or it may be felt by someone who is unhappy or bored. Fatigue usually disappears after rest and relaxation. A feeling of being very tired all the time, or **chronic** fatigue, is usually a **symptom** of an illness.
After the race, he was overcome by fatigue.

fatty acid *noun*
Fatty acid is a substance which makes up fat. Fatty acids are needed to help the body produce **energy** and make use of important **vitamins**, such as A, D, E and K. Some kinds of fatty acid are described as essential. This means that the body cannot make them. This is why fat is one of the important parts of a healthy **diet**.
She lost weight by cutting down on fatty acids in her diet.

feces *noun*
Feces are a **waste product** found in the **intestine**. They are also called stools or bowel movements. Feces are what is left of food after all the **nutrients** have been broken down, or **digested** and absorbed into the body. They are formed in the large intestine and passed out through the **anus**. Normal feces are brown and soft.
Feces contain harmful bacteria so it is important to wash the hands after going to the toilet.

feelings *noun*
Feelings are different moods, or states of mind. They are also called emotions. Love, hate, happiness and sadness are all feelings. Some feelings have an effect on the body. Anxiety, fear or excitement make the heart beat faster and breathing becomes more rapid. **Hormones** have an effect on feelings. During adolescence, the hormones are working hard to help a child grow into an adult. This may give a person very strong feelings.
She was so angry, she gave way to her feelings and shouted.

female *adjective*
Female describes anything to do with girls or women. A female child is a girl. The opposite of female is **male**.
Boys were not allowed in the female changing rooms.

femur *noun*
A femur is a **bone** which joins the **hip** to the
knee. The femur, or thighbone, is the longest
and one of the strongest bones. Powerful
muscles help the femur move.
The femoral artery runs down the thigh.
femoral *adjective*

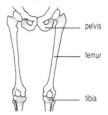

pelvis

femur

tibia

fertile *adjective*
Fertile describes a person who is able to
have children. Fertile women have **ova** that
can be fertilized by **sperm**. A fertile woman
must also have a **uterus** that can carry a
baby. Fertile men have sperm that are able
to fertilize eggs. Women can be fertile from
puberty to about 45 years of age. Men can
be fertile from puberty until old age.
*The tests showed that she was fertile and
could have a baby.*
fertility *noun*

fertilization *noun*
Fertilization describes what happens when a
sperm meets an **ovum**. Fertilization is also
called **conception**. The sperm and ovum join
and turn into a bundle of **cells** called a
zygote. This grows into an **embryo**, and then
a **fetus**.
*Fertilization of a woman's ovum can be
artificially carried out in a laboratory.*

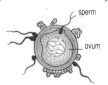

sperm

ovum

fetus *noun*
A fetus is a baby in the **uterus**. An unborn
baby is called a fetus from two months after
conception until it is born. The fetus floats in
amniotic fluid and receives its nourishment
through the **umbilical cord**. For the first two
months an unborn baby is called an **embryo**.
*The doctor used an ultrasound scanner to
examine the fetus.*

fever *noun*
A fever is a rise in body heat. The level of
heat in a body is called the **temperature**.
When the temperature of a person rises
above the normal level they have a fever.
A person with a fever may feel hot and then
cold and shivery. They may feel tired and
have a **headache**. Fevers are usually a
symptom of an **infection**. Fevers can be a
sign of serious illness.
*She stayed in bed suffering from a high
fever.*

fibre *noun*
Fibre is the layer of tiny strands that makes
up **cell** walls in plants. Vegetables, fruits,
beans and cereals all contain fibre. Fibre
cannot be broken down, or **digested**. As it
travels through the **large intestine**, fibre
absorbs water and forms **feces**. People who
do not eat enough fibre suffer from
constipation.
*He ate plenty of fibre in his diet to stay
healthy.*
fibrous *adjective*

fibroid *noun*
A fibroid is a lump of **tissue**. It is formed from
tiny strands of fibre. Fibroids sometimes
develop in the **uterus**. They are a harmless,
or **benign**, growth and often cause no
problems. Large fibroids may cause cramp-
like pains or heavy bleeding during
menstruation. They may also stop a woman
having a child. Fibroids that cause problems
can be removed by surgery.
*She recovered quickly after the fibroids were
removed.*

fibula *noun*
A fibula is the long, thin **bone** that runs down the outside of the lower leg. The top end of the fibula joins the shinbone, or **tibia**, just below the knee. The bottom end forms the outer side of the ankle joint. **Muscles** attached to the fibula help the leg to move.
The boy fractured his fibula just above the ankle.

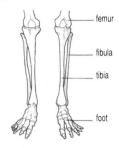

femur

fibula

tibia

foot

finger *noun*
A finger is part of a **hand**. Fingers, or digits, are found at the end of each hand. Four of the fingers on each hand are made of three bones called **phalanges**. The two fingers called thumbs have only two phalanges. The bones of the fingers are joined by **joints** that act like hinges. **Tendons** along the fingers move the joints. These movements are controlled by muscles in the forearm. Many **nerve endings** in the fingers make them very sensitive to touch.
There are four fingers on each hand.

fingernail *noun*
A fingernail is a piece of hard **tissue**. It is made of a horny substance called **keratin** which is waterproof. Fingernails grow at the end of each finger from the root of the nail. The nail itself is not supplied with **blood** or **nerves**. It is dead tissue and does not hurt when it is cut.
She bit her fingernails, so that the fleshy ends of her fingers were not protected.

first aid ► page 60

flex *verb*
To flex is to bend. **Joints** allow people to flex their body.
She flexed her biceps muscle when she lifted the jug of water.
flexible *adjective*

fluoride *noun*
Fluoride is a chemical substance. It is found naturally in soil and water. Fluoride helps **bones** and **teeth** to form and helps to prevent tooth decay, or **caries**. Taking too much fluoride can be harmful to teeth.
Fluoride is often added to water supplies and toothpaste to help keep people's teeth healthy.

follicle ► **hair follicle**

food *noun*
Food is all the substances that nourish the body. Plants and animals supply all the food the body needs. Food can be arranged in four groups. These are a milk group, a meat group, a bread and cereal group and a fruit and vegetable group. A healthy **diet** is a balance of these foods. Vegetarians, or people who do not eat meat, can still eat a healthy diet from other kinds of food. Food is measured in units of energy called **calories**.
Food helps the body to grow, keep healthy and heal itself.

food poisoning *noun*
Food poisoning is an illness. It attacks the **digestive system** and causes **vomiting**, **diarrhea** and stomach pains. Food poisoning is usually caused by eating food that contains harmful **bacteria**. Food that contains poisonous chemicals may also cause food poisoning. Some plants, such as some fungi, are very poisonous to humans and should never be eaten. People recover from most attacks of mild food poisoning after a day or so.
He took special fluids to treat his food poisoning.

first aid *noun*

First aid is immediate treatment that is given to a person who is taken ill or has had an accident. First aid can save a person's life. It can also prevent a person's condition getting worse. Someone should call for a doctor or other help as quickly as possible.
On the first aid course, he learned how to tie a sling for a broken arm.

Someone who feels faint should sit down, with the head between the knees.

Burns should be cooled under running water.

If someone is not breathing, artificial respiration should be given.

If poisoning is suspected, an ambulance should be called immediately.

A person in shock should lie down, with the feet raised slightly. Tight clothing should be loosened.

foot (plural **feet**) *noun*
A foot is a part of the body that lies at the
end of the lower leg. It is made up of 26
small **bones**. Each foot has a **heel**, an **arch**
and five **toes**. The arch of the foot is formed
by five bones called the metatarsals that join
the ankle to the toes. The largest bone in the
foot is the heel bone, or **calcaneus**.
*Feet are used for standing, walking and
running.*

forceps *plural noun*
Forceps are pincers used for holding and
seizing objects in surgical **operations**.
They had to use forceps to deliver the baby.

forehead *noun*
A forehead is part of a **face**. It is the front
part of the **head** above the **eyes**. The
forehead, or brow, runs from the hairline to
the **eyebrows**.
Her long hair fell down over her forehead.

foreskin *noun*
A foreskin is part of a **penis**. It is a fold of
skin that covers the end of the penis. The
foreskin can be pulled back in boys after the
age of about three years. Before this it is
held in place by fine strands of **tissue**.
Cutting away part or all of the foreskin is
called **circumcision**.
It is important to keep the foreskin clean.

fracture ► page 62

freckle *noun*
A freckle is a small brown patch on the skin.
Freckles appear on some people when they
are in sunlight. This is because the body
makes more **melanin** to protect the skin from
burning. Melanin is the pigment which makes
skin darker. Fair-skinned people are more
likely to have freckles than dark-skinned
people. Freckles fade when the skin is
covered or if the weather is cloudy.
*After a day in the sun, freckles appeared all
over her nose.*

fungus (plural **fungi**) *noun*
A fungus is a plant. Some tiny fungi can live
and spread on humans, usually on the skin.
Fungi are sometimes breathed into the **lungs**
and develop into an illness. Fungal infections
include **athlete's foot** and **ringworm**.
*Fungicide is a substance that kills fungi and
can be used to treat some fungal infections.*

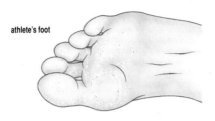
athlete's foot

funny bone ► **humerus**

fracture *noun*

A fracture is a broken **bone**. Fractures are one of the most common injuries that result from accidents. The treatment of a fracture is usually straightforward. A broken limb is often set in a **plaster cast** so that it cannot move until the break heals. The bones of older people are more brittle than those of younger people. They break more easily and take longer to heal.

The X-ray showed that he had a fracture of the tibia.

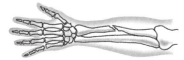

Simple fracture. A clean break.

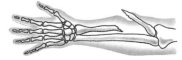

Open fracture. The bone pierces through the skin.

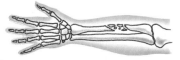

Comminuted fracture. The bone is shattered at the break.

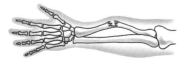

Greenstick fracture. The bone breaks only on one side. This kind of fracture is common among children.

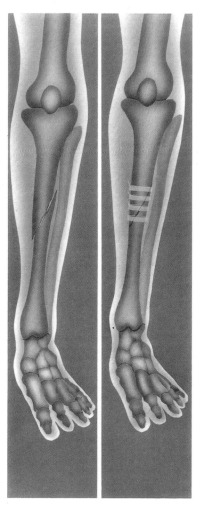

A metal pin, plate, screws or wire may be inserted where a fracture is very difficult to treat.

gall bladder   *noun*
A gall bladder is an **organ** in the body. It is
about 10 centimetres long and shaped like a
pear. The gall bladder lies under the right
side of the **liver**. It is attached to the liver and
the **duodenum** by a tube called the **bile**
duct. Bile is a digestive juice made by the
liver and stored in the gall bladder. From
here it is passed along the bile duct to the
intestines.
*Some people develop small lumps or stones
in their gall bladder.*

gastric   *adjective*
Gastric describes anything to do with the
stomach. Gastric juice is the **acid** liquid the
stomach makes to break down, or **digest**,
food.
*His gastric problems caused him to have
indigestion.*

gastroenteritis   *noun*
Gastroenteritis is an **inflammation** of the
stomach and intestine. People with this
illness have **diarrhea**, stomach pains and
vomiting. If the symptoms are severe, a
person may suffer from loss of fluid, or
dehydration. Dehydration can be dangerous
in babies and children, and they may need to
be treated in hospital. Gastroenteritis is
caused by harmful **bacteria** in food, **viruses**,
allergies or poisons.
*Someone with gastroenteritis should rest and
drink plenty of fluids.*

gene ► **genetics**

genetics ► page 64

generation   *noun*
Generation refers to an age or period of time.
People of the same age are of the same
generation. The children of those people are
the next generation. **Characteristics** of
people are passed from one generation to
the next. Every child born starts a new
generation. Generations of people are
counted in periods of about 30 years.
*Her parents belong to a different generation
from her.*

germ   *noun*
A germ is a tiny living thing. Germs are too
small to be seen by the human eye, but they
can be seen under a microscope. **Bacteria**,
viruses and **fungi** are all germs. Some
germs are harmless. Others only cause
diseases when they enter the body and start
to grow in number, or multiply.
*When he had a cold, he tried not to spread
germs by covering his mouth when he
coughed.*

German measles ► **rubella**

gestation ► **pregnancy**

gingivitis   *noun*
Gingivitis is an **inflammation** of the **gums**.
The gums become sore and swollen. They
bleed when the teeth are brushed. The
inflammation is caused by harmful **bacteria**
in the mouth that form a hard, scaly
substance called **plaque** at the base of each
tooth. Plaque irritates the gums and causes
the inflammation of gingivitis.
*Gingivitis may result from not brushing the
teeth properly.*

gland   *noun*
A gland is an **organ** of the body. Glands
make and store different kinds of fluid, such
as **sweat**. There are two kinds of gland in the
body. Exocrine glands pass fluids along
tubes called **ducts**. **Endocrine glands** pass
fluids straight into the **bloodstream**.
The lymph glands in her neck are swollen.

genetics   *noun*

Genetics is the study of how features are passed from parents to children through their **genes**. The **nucleus** of each human body cell contains 46 **chromosomes**. Every chromosome carries **genes** that contain **hereditary** information. Physical **characteristics**, some mental abilities, and hereditary disorders are inherited by children from their parents.

Genetics has helped scientists to produce useful medicines cheaply and easily.

Boy or girl

An egg fertilized by a sperm with an x-chromosome will become a girl.

Chromosomes decide a person's sex. The father produces two kinds of sperm. One kind has a special x-chromosome, and another has a y-chromosome. The mother's egg cell has only an x-chromosome.

An egg fertilized by a sperm with a y-chromosome will become a boy.

Strong and weak genes

Human body cells contain 23 pairs of chromosomes. One of each pair is inherited from the mother and one from the father. If a boy inherits a brown-eye gene from his mother and a blue-eye gene from his father, the boy will have brown eyes, because the brown-eye gene is stronger, or dominant. But the boy will still have a 'hidden' blue-eye gene. When he grows up and has a child himself, he might pass on his blue-eye gene to one of his children.

Growing cells

Chromosomes are made of a protein called DNA. Each chromosome carries about 1,000 genes that are strung together in a long, twisted chain. When body cells multiply, they divide in two. As this happens, the chain of chromosomes makes an exact copy of itself, one for each new cell. So each body cell contains exactly the same genetic information.

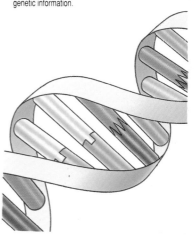

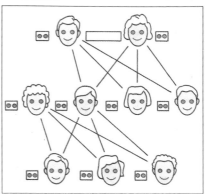

glandular fever *noun*

Glandular fever is a **disease** caused by a **virus**. It is most often caught by young people between 15 and 25 years. The first **symptoms** are feeling tired and slight **fever**. This lasts for a week to 10 days. A sore throat and high fever then develop. The glands in the neck swell and there may be a faint, pink **rash** over the body. This part of the illness lasts for a further week to 10 days.
Glandular fever is often called the 'kissing disease'.

glaucoma *noun*

Glaucoma is a **disease** of the **eye**. It may develop suddenly, or build up slowly. Glaucoma is caused by a build-up of fluid in the centre of the **eyeball**. The fluid presses against the back of the eye and the **retina** becomes damaged. Sudden, or **acute**, glaucoma causes pain in the eye and vision is blurred. A person with glaucoma is unable to see things around the object they are looking at. This is called tunnel vision.
Glaucoma can be helped by drugs which get rid of the extra fluid.

glottis ► **larynx**

glucose *noun*

Glucose is a kind of sugar. It is found in most foods that contain sugar and starch. People need glucose for **energy**. Glucose is absorbed straight into the **bloodstream** from the **intestine**. The brain needs glucose in order to function properly. **Hormones** in the body balance the levels of glucose.
Glucose which is not used up as energy is stored as fat.

goitre *noun*

A goitre is a swelling of the **thyroid gland**. It is usually caused by a lack of iodine in the diet. A goitre can also result when the **pituitary gland** works too hard, or is overactive.
A goitre can be treated by adding salt with iodine to a person's diet.

graft *noun*

A graft is an **operation** to repair or replace a part of the body. Healthy **tissue** is taken from a different part of the body, or another person. It is then attached to the damaged area and left to heal. Skin is the most common part of the body used to make a graft. Grafts can also be made with whole organs, such as hearts, kidneys and livers. Bones, muscles, tendons and many other parts of the body are also used as grafts.
A skin graft is a very useful treatment for severe burns.

graze *noun*

A graze is a small **wound** on the **skin**. It is caused when the skin is scraped or rubbed hard enough to make it bleed. A graze should be gently cleaned and any dirt or gravel removed. A **scab** forms over the wound and falls off as the skin heals.
Children often graze their knees when they fall over.

grey matter ► **cerebral cortex**

growth ► page 66

gullet ► **esophagus**

gum *noun*

A gum is part of the **mouth**. Gums are the pink, dense **tissue** in which **teeth** are buried. A moist layer of **skin** covers the gums.
Teeth and gums should be brushed at least once a day to keep them healthy.

gut ► **intestine**

gynecology *noun*

Gynecology is the study and treatment of illness that is only found in girls and women. Gynecology covers **menstruation**, and also deals with any problems to do with the **breasts** and genitals.
There is a special gynecology unit in most hospitals.

growth *noun*

1. Growth is an increase in size. It also means development. The growth of a human **embryo** into an adult takes many years. Most people stop growing when they are between 18 and 24 years old. Growth is controlled by a **hormone** in the **pituitary gland** of the brain.
Foods containing proteins, vitamins and minerals are essential for healthy growth.
2. A lump in the body, or **tumour**, is sometimes called a growth.
The growth on her neck had to be surgically removed.

As a person grows, the body gets bigger in relation to the head.

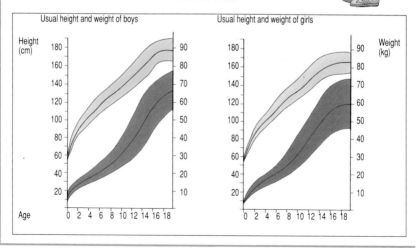

Usual height and weight of boys

Usual height and weight of girls

Height (cm)

Weight (kg)

Age

The pituitary gland in the brain produces the growth hormone. The hormone is released into the bloodstream and stimulates growth in the body.

Hormones produced by the thyroid influence cell growth.

Hormones produced by the pancreas also influence cell growth.

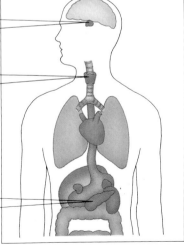

hair ► page 69

hair follicle *noun*
A hair follicle is a tiny space under the **skin** that produces a **hair**. The root of a hair is planted at the bottom of the follicle. Oily fluid passes into the follicle from the **sebaceous gland** beside it. This protects the hair. A tiny **muscle** attached to the follicle can make hairs stand on end.
Hair follicles stand up when a person is cold or afraid and give them goose pimples.

halitosis *noun*
Halitosis is bad or unpleasant-smelling breath. It is usually caused by not cleaning the teeth and gums properly, or it may be a sign of a throat, nose or lung **infection**. Sometimes it is a symptom of other illnesses.
Halitosis can sometimes be cured by a change of diet.

hammer *noun*
A hammer is one of the three tiny **bones** in the **middle ear**. The other two bones are called the **anvil** and the **stirrup**. Together, they are called the **ossicles**.
The hammer is a bone that helps in the process of hearing.

hand *noun*
A hand is the end part of the arm. Each hand has five nails at the end of five fingers. The palm of each hand is formed by five bones called metacarpals which join the wrist to the **fingers**. The whole hand has 27 bones and a web of **muscles** and **tendons**.
The hands can move in many different ways.

hay fever *noun*
Hay fever is an **allergy**. It is a reaction to tiny particles in the air, such as dust or pollen from plants. People with hay fever feel as if they have a cold. They cough and sneeze, have a runny, stuffy nose and watering eyes. Hay fever is **hereditary**, which means it can be passed down in certain families. There are drugs which help the symptoms of hay fever.
She always suffers from hay fever in the hot weather.

head *noun*
1. The head is the top part of the body. The head contains the **brain** and includes the **face**, **scalp** and **ears**. It is formed by the bones of the **skull**.
She covered her head with a hat to keep warm.
2. A head is the rounded end of a **bone** which fits into another to form a **joint**.
The head of the thighbone, or femur, fits into the pelvis.

headache *noun*
A headache is a pain in the **head**. It can be across the **forehead** or deep inside part or all of the head. Headaches are most often caused by feeling tired or upset. They can be caused by an **infection**, a disorder of the **eye** or **ear**, or a head injury. Some headaches may be a **symptom** of another illness.
She took a painkilling drug to treat her headache.

heal *verb*
Heal describes the way the body mends itself when it is ill or injured. To heal is to make well. Some **cells** renew themselves to repair damage, such as when the skin is cut. Other cells fight **germs** and allow the body to recover from **infection**. The body's **immune system** is important in helping the body to heal. **Medicines** and **operations** also help to heal the body.
His graze healed within a few days.

hair *noun*

Hair is a part of the body that grows from the **skin**. Each hair is a fine thread of tough **protein** called **keratin**. Keratin cells are not alive. Hair grows all over the body except for the palms of the hands and the soles of the feet. The thickest area of hair grows from the **head**. Hairs grow, fall out and are replaced by new ones all the time. Hairs on the head last from two to five years before they are shed. Hair protects the body from dust and dirt and helps to keep it warm.

There are about 100,000 hairs on a human head.

Hair colour is decided by a pigment called melanin in the hair shaft. Older people's hair often turns grey or white because melanin stops being produced.

Hair texture depends on the shape of the hair follicles.

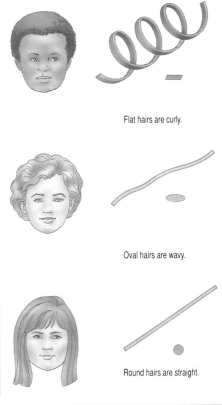

surface of skin

shaft containing melanin

sebaceous gland

muscle

hair bulb

follicle — root

papilla

blood vessel

Flat hairs are curly.

Oval hairs are wavy.

Round hairs are straight.

health *noun*

Health is the condition of body and mind.
People with good health are fit and well.
People with bad health may be suffering
from some kind of illness.
Everyone feels better if their health is good.

Eating a balanced diet keeps people healthy.

Exercise strengthens the muscles and improves breathing
and blood circulation.

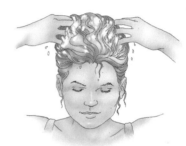

Keeping clean controls the growth of harmful bacteria.

Good personal relationships are important.

People who do not get enough sleep do not function well.

health ► page 70

hearing   *noun*
Hearing is being able to listen to sound. The
ear is the organ of hearing. The opposite of
hearing is **deafness**.
*A person who is hard of hearing may use a
hearing aid.*

heart ► page 72

heartbeat   *noun*
A heartbeat is the pumping action of the
heart. A heartbeat is felt every time the
muscles of the heart contract, to push **blood**
round the body. The heart beats about 72
times a minute in an adult at rest. Heartbeats
can be felt as a throb of pressure, or **pulse**,
in the **arteries** of the wrist and neck.
*The heartbeat of a young child is faster than
that of an adult.*

heart disease   *noun*
Heart disease is any disorder that affects the
heart. Any kind of heart disease causes
problems with the **circulation** of **blood**.
Having an **operation** can cure some kinds of
heart disease. Other kinds can be helped by
drugs. **Rheumatic fever**, **angina** and
hypertension are examples of heart
disease. Coronary heart disease is caused
when the heart muscle does not get enough
blood.
*Eating less fat, exercise and not smoking
cuts down the risk of heart disease.*

heat   *noun*
Heat can cause illness. If the body becomes
too hot, a person may have heat exhaustion
or heatstroke. They may sweat and shiver,
feel dizzy, sick and have a headache. The
body is unable to control its temperature.
A person may develop dry, hot skin and
become very confused. **Shock**,
convulsions, **coma** and even death may
follow without emergency treatment.
*People with heat exhaustion should rest
quietly in a cool place, and drink some water.*

heel   *noun*
A heel is the rear end of the foot. It is formed
by the heel bone, or **calcaneus**, resting on a
thick pad of skin. Heels bear the full weight of
the body and act as shock absorbers.
*The correct way of walking is to put the heel
down first.*

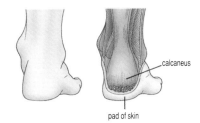

calcaneus

pad of skin

hemoglobin   *noun*
Hemoglobin is the substance in **red blood
cells** which colours the cells red. Its function
is to carry **oxygen** from the **lungs** to the rest
of the body.
*Sickle-cell anemia is a disease caused by an
abnormal hemoglobin.*

hemorrhage   *noun*
A hemorrhage is **bleeding** from a damaged
blood vessel. This can happen inside or
outside the body. A heavy hemorrhage is
dangerous and must be stopped as quickly
as possible.
*Pressing firmly on a wound helps to stop a
hemorrhage.*

hemorrhoids   *plural noun*
Hemorrhoids are a disorder of the **anus** and
rectum. They are a bunch of swollen veins
that bulge either inside, or just outside the
anus. They cause pain, burning and itching.
They may bleed when **feces** are passed.
Constipation or **pregnancy** sometimes
causes hemorrhoids, but they can appear
and disappear at any age.
*Hemorrhoids are more commonly known as
piles.*

heart *noun*

The heart is an organ made of **muscle**. It is about the size of a clenched fist, and lies between the **lungs**. The heart is divided into four sections, or **chambers**. These are the left and right **atrium**, and the left and right **ventricle**. The heart pumps **blood** around the circulatory system. Blood carries essential **oxygen** and **nutrients** to each **cell** in the body. The body cannot live for more than about five minutes if the heart stops completely.

The heart of an adult beats about 70 times a minute.

Valves in the heart make sure the blood flows the correct way. Many large veins also have valves.

valve open

valve closed

How the heart works

pulmonary artery carries blood to lungs

vena cava carries blood from the body's cells

aorta carries blood to the body's cells

pulmonary veins carry blood from lungs

left atrium

valve

left ventricle

right atrium   right ventricle

The heart relaxes. Blood flows in to both atria from the veins.

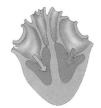

The atria contract, and push blood through the valves into the ventricles.

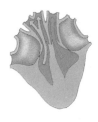

The ventricles contract, pushing open the semilunar valves. Blood flows into the arteries. This contraction is felt as a heartbeat.

hepatitis *noun*
Hepatitis is an **inflammation** of the **liver**.
When something is wrong with the liver,
there is pain and sometimes **jaundice**
develops. Hepatitis may last for six months
or more. It is usually caused by four kinds of
virus called A, B, C and D. They are all
infectious. Type A is caught from water or
food which has been contaminated with
feces or **urine**. The other types are caught
by contact with other body fluids, such as
blood.
He caught hepatitis while on holiday in India.

heredity *noun*
Heredity describes how parents pass on
characteristics to their children. These can
be to do with the body, such as a family
likeness. Or they may be to do with the brain,
such as being good at mathematics at school.
Sometimes it is a tendency towards a certain
illness. A feature that can be traced back
through the family is described as hereditary.
Genetics is the study of heredity.
hereditary *adjective*

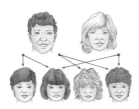

herpes *noun*
Herpes is a skin **infection** caused by a
virus. A small blister appears and then
breaks into a sore. The sore then dries up,
heals and forms a **scab**. Herpes around the
mouth and nose are usually called cold sores
because they often appear with a cold.
Another kind of herpes appears on the
genitals. Both kinds of herpes tend to come
back again and again.
*Sometimes there is a long time between
outbreaks of herpes, but there is no cure.*

hiccup *noun*
A hiccup is a short breath of air sucked in
very suddenly. This happens when the
diaphragm muscle in the stomach tightens
in a spasm. Air is pulled in and hits the
epiglottis, the flap that covers the **larynx**.
The air rushes in so quickly it is like banging
a tiny drum. This causes the 'hic' noise of a
hiccup.
*An upset stomach or stress are two reasons
for hiccups.*

hinge joint ► joint

hip *noun*
A hip is a part the body which is found each
side of the **pelvis**, where the legs join the
rest of the body. The **femur** is attached to the
pelvis by a ball and socket **joint**. The pelvis
is formed from the two hipbones and the
lower spine.
*Women usually have wider hips than men,
which helps them during childbirth.*

HIV *noun*
HIV is the **virus** which causes **AIDS**. HIV
stands for Human Immunodeficiency Virus.
It slowly attacks the body's defence against
illness, or **immune system**. HIV is caught by
coming into contact with infected body fluids,
such as blood or semen. Any sexual contact
carries the risk of HIV infection. People with
HIV may not feel ill or know they are carrying
the virus. It may be many years before the
disease of AIDS develops. A blood test can
show if a person is infected.
There is no known cure for HIV infection.

hives *noun*
Hives are red, painful swellings on the **skin**.
They are usually caused by an unusual, or
allergic, reaction to a drug or an insect bite.
Hives can also be an allergic reaction to
certain foods, such as shellfish. Hives can
appear anywhere on the body and last a few
days. There are **drugs** which ease the
symptoms.
The bee sting gave her hives.

hygiene *noun*

Hygiene is a way of staying healthy by keeping clean. Bodies need to be washed and teeth brushed to stop **bacteria** building up. Most bacteria on the body are harmless, but some kinds, such as in **feces**, can cause illness. It is important always to wash the hands after using the toilet and before touching food.

Public hygiene is also very important in keeping people healthy. It involves steps taken by governments and voluntary health organizations to improve the general health of a community. Sewage systems and immunization programmes are part of public hygiene.

Personal hygiene such as washing your hands regularly is as important as public hygiene.

Brushing teeth keeps them healthy. Without regular brushing, bacteria in the mouth combine with food particles and acid to form plaque. Plaque causes tooth decay and gum disease.

Minor wounds should be washed carefully then covered with a sterile dressing to stop bacteria getting into the cut and causing an infection.

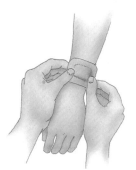

Washing hands destroys harmful bacteria, which could cause illness.

Flies cause disease by spreading bacteria. Food should be covered or kept in the refrigerator to prevent flies from coming into contact with it.

Careful preparation of food prevents growth of bacteria. Rules of hygiene such as keeping hands clean and touching prepared food as little as possible must be followed.

Careful control of the way food is processed makes sure that it is safe to eat by destroying any bacteria that may be present.

Sewage systems process waste hygienically, to keep harmful bacteria away from public places.

homeopathy *noun*
Homeopathy is a kind of **alternative medicine**. It is a way of treating illness with small amounts of special medicines. These are mostly made of herbs, mixed with alcohol and diluted with water.
Treatment by homeopathy may take a long time.

hookworm *noun*
A hookworm is a worm that is usually found in tropical countries. Hookworms get into the body through the **skin**, usually on bare feet. They find their way to the **small intestine** where they live and grow up to 12 millimetres long. Hookworms attach themselves to the wall of the small intestine and suck blood for food. People with hookworm feel sick, have **diarrhea** and stomach pain.
People with hookworm may lose blood and develop anemia.

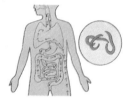

hormone *noun*
A hormone is a chemical substance made in the body. **Organs** such as the **endocrine glands** make different kinds of hormone.
Hormones make the body develop and work properly.

hospice *noun*
A hospice is the word for a place that looks after dying people. Hospice is also a kind of medical care for people who are close to **death**. People with a serious illness that cannot be cured need special care. Old people who are coming to the end of life also need this kind of care. Hospice care gives nursing and support for the patient, their family and friends.
She visited her aunt at the hospice.

hospital *noun*
A hospital is a place where people go to have medical treatment they cannot have at home, or through their family doctor.
Patients stay in the hospital for a few hours or longer for some kinds of treatment, such as an **operation**.
Hospitals deal with all kinds of disorder of the human body and mind.

host *noun*
A host is an animal whose body provides food and a home for another animal, such as a **tapeworm**.
Animals which exist in or on living hosts are called parasites.

human *noun*
A human is any one of the human race. Humans are mammals, or animals with warm blood which feed their babies with breast milk. Humans are the most powerful species on Earth. This means that we control and use most other life forms through our much greater intelligence.
Young humans need to be cared for longer than any other animal.

humerus *noun*
A humerus is the **bone** of the upper arm. It joins the shoulder to the elbow. The lower end of the humerus is sometimes called the funny bone. Hitting the back of the elbow often strikes the ulnar nerve on the funny bone. This causes a tingling pain which shoots up and down the arm.
It was not funny when he broke his humerus.

hunger *noun*
Hunger is the feeling of wanting to eat food. People suffering from hunger sometimes have a pain in the upper part of the **abdomen** and the **stomach** may gurgle and rumble. The sight or smell of food makes saliva rush to the mouth of a hungry person. Hunger is controlled in a part of the brain called the **hypothalamus**.
Hunger made her feel empty and weak.

hygiene ► page 74

hypertension *noun*
Hypertension is high **blood pressure**.
Very high blood pressure may cause
headaches, but often there are no
symptoms. Hypertension is usually
discovered at a **check-up** when blood
pressure is taken. Being overweight, eating
too much salt, drinking too much alcohol and
suffering **stress** may be some of the reasons
for hypertension.
*Hypertension can damage the heart, blood
vessels and kidneys unless it is treated.*

hypothalamus *noun*
The hypothalamus is part of the **brain**.
It links the **endocrine system** and the
nervous system in the brain stem. The
hypothalamus lies below the **thalamus** and
above the **pituitary gland**.
*The hypothalamus controls appetite, thirst
and body temperature.*

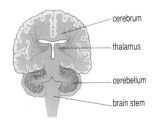

- cerebrum
- thalamus
- cerebellum
- brain stem

hypothermia *noun*
Hypothermia is an illness caused by a
person becoming very cold. The
temperature of the body drops to below
normal. All the **organs** in the body slow
down and it is difficult to move. The person
becomes confused and cannot talk properly.
Someone with hypothermia may not realize
they are in danger. They may then become
unconscious, fall into a **coma** and die.
*People who are not dressed for cold weather
are at risk of hypothermia.*

ileum *noun*
The ileum is the lowest part of the **small
intestine**. In an adult, it is about 3.6 metres
long. The ileum runs from the middle part of
the small intestine, or jejunum, to the start of
the **large intestine**. Digested food is
absorbed into the blood in the ileum.
*Undigested food passes from the ileum into
the large intestine.*

ilium (plural **ilia**) *noun*
An ilium is a bone in the **pelvis**. There are
two ilia, which are the largest upper parts of
the hipbone on each side. The ilia are fused
to two other bones called the **pubis** and the
ischium.
*Strong muscles in the upper legs and
buttocks are attached to the ilia.*

illness *noun*
An illness is a sickness or **disease**. It is a
state of being unwell in the body or mind. An
illness may be mild or serious, have a
name or be unknown.
Her illness was difficult to diagnose.

immature *adjective*
Immature describes someone who is not fully
grown. An immature person is not **adult**.
Immature can also describe someone's
personality and behaviour. The opposite of
immature is **mature**.
*The baby held on to the chair because he
was too immature to stand alone without any
help.*
immaturity *noun*

immune system ► page 78

immune system   *noun*

The immune system is the body's defence against **illness**. The main parts of the system are white **blood cells** which are made in the **bone marrow**, **lymph nodes** and **spleen**. When harmful **bacteria** or **viruses** enter the body, the white blood cells go to the site of the infection or injury. Some white cells called phagocytes eat the bacteria. Others called lymphocytes make **antibodies** which help to fight the infection. *The immune system prevents illness by developing a resistance to some germs.*

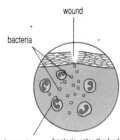

bacteria enter the body

phagocytes crowd round the bacteria

phagocytes surround the bacteria and destroy them

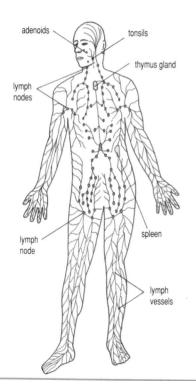

Lymph is produced in the adenoids, the tonsils, the thymus gland, the spleen and the lymph nodes. Lymph circulates around the body through thin-walled vessels.

The body needs extra lymphocytes to fight an infection. So lymph nodes in the neck, armpits and groin swell up as they produce more lymphocytes.

immunity *noun*

Immunity describes the way the body fights off an illness. The **white blood cells** in the **immune system** remember how to fight an illness if it comes back. This stops a person having the illness again. Because of this, many diseases such as glandular fever can only be caught once. People also have immunity to certain diseases after they are given a **vaccination**. Babies receive some immunity from their mothers, passed across the **placenta** and in **colostrum** after birth.
Few people develop immunity to the common cold.

immunization *noun*

Immunization is a way in which the body is helped to fight off an **infection**. This is done by giving **vaccinations**. The medicines used for immunization are given in a course of **injections**, or by mouth. The immunization of babies protects them against many diseases, such as **measles** and **mumps**.
Immunization prevented him catching whooping cough.

impetigo *noun*

Impetigo is a skin **disease.** It starts as a red patch which turns into small blisters. The blisters break and form crusty, yellow sores that ooze and spread over the body. Impetigo is caused by **bacteria**. It can develop from another skin disease, such as **eczema**. Impetigo is very easy to catch, or **contagious**, especially between children.
Antibiotic creams and drugs help to clear up impetigo.

implant *noun*

An implant is something placed in or under the skin. This may be **tissue** from another part of the body or a **drug**. It can sometimes be a pellet filled with a **hormone**, or a tube which feeds a substance into the body.
The hormone implant stopped her becoming pregnant.

incisor ► **tooth**

incubator *noun*

1. An incubator is a piece of hospital equipment. It is a special container for babies that are born very small, or sick. The container is a clear box that protects the baby from **infection** and keeps it warm. The air inside has the right mixture of moisture and oxygen for a tiny baby.
The premature baby was placed in an incubator.

2. Other kinds of incubator are used in medical tests.
The incubator provided the right mixture of air, moisture and warmth to grow bacteria.

indigestion *noun*

Indigestion is not being able to digest food properly. **Symptoms** may be stomach pain and feeling sick. Indigestion may also cause a burning pain in the chest. This kind of pain is called heartburn and is caused by acid in the stomach rising up. Indigestion may occasionally be a symptom of a more serious illness.
Indigestion is most often caused by eating too much too fast.

infection *noun*

An infection is an illness which is caused by **germs** entering the body. **Viruses**, **bacteria** and worms all cause infections. Germs can be breathed in, taken in food or water, or passed through the skin. Some infections can be treated with drugs called **antibiotics**.
People and animals can pass on infections.

infectious disease *noun*
An infectious disease is an **illness** that can be passed from one person to another. There is a cure or a **vaccine** for many infectious diseases. These diseases include **diphtheria**, **measles** and **mumps**.
An infectious disease may be passed to another person.

infertility *noun*
Infertility is not being able to have children. Sometimes it happens because the woman's **ovaries** or **uterus** are not working properly. It may be caused by the man not making enough, or the right kind of, **sperm**. Often a cause cannot be found. Infertility can last a short time, or for ever.
Some kinds of infertility can be cured.
infertile *adjective*

inflammation *noun*
Inflammation is the reaction of the body to an **injury** or **disease.** The part of the body affected by inflammation becomes painful, red and swollen.
Inflammation disappears as the body starts to heal.

inflammation of an
injured finger

influenza *noun*
Influenza is an **infectious disease** caused by a **virus.** After catching influenza, a person will feel shivery and unwell. A headache, high temperature and sore throat develop. People with influenza feel weak, their muscles and joints ache and they have a dry, painful cough.
Influenza usually lasts about five days, but a person may feel tired for a while afterwards.

inhale *verb*
To inhale is to breathe in. Some illnesses can be treated by inhaling certain **drugs** or gases. These make it easier for a person to breathe. People with **asthma** often carry an inhaler. This is a small container that puffs a mist of drugs into the mouth to be inhaled.
He inhaled the medicine once a day.
inhalation *noun*

injection *noun*
An injection is a way of putting fluid into the body. This is done by using a needle attached to a tube of fluid called a **syringe**. Injections can be put into the skin, or through the skin and into a muscle or a vein. They can also be put into a **joint** or around the nerves of the spinal cord.
Some diabetics give themselves an injection of insulin daily.
inject *verb*

inner ear *noun*
The inner ear is the innermost part of the **ear**. It sends messages of **hearing** and **balance** to the brain. The **cochlea** in the inner ear is a fluid-filled tube, coiled like a snail shell, that passes impulses to the **auditory** nerve. The inner ear also has three more tubes called the **semi-circular canals**. Fluid in these tubes moves as the head moves. This information is sent to the brain. The brain then tells the rest of the body how to move.
The inner ear helps people to keep their balance.

inoculation ► vaccination

insomnia *noun*
Insomnia is difficulty in sleeping. There are many different causes of insomnia, such as illness, pain, heat, noise and depression.
Anxiety about his exams caused his insomnia.

instinct *noun*
Instinct is a kind of **behaviour** that is not learned. It is an action that a person does without thinking, such as holding out their arms to break a fall. Instinct makes the body move quickly before there is time to think. This is usually when some kind of danger threatens. Instinct helps people to survive.
New-born babies turn their head towards the breast and suck by instinct.

insulation *noun*
Insulation is protection. In the body, the layers of **skin** and **fat** act as insulation for the inside of the body. This kind of insulation holds in the heat and helps to keep the body at the right **temperature.** The outer layers of skin also prevent harmful substances or objects entering the body and damaging the **organs**.
A person does not need to be overweight to have enough insulation from the cold.

insulin *noun*
Insulin is a substance called a **hormone**, which is produced by the body. It is made in the **pancreas** and released into the **bloodstream.** Insulin controls the levels of blood sugar, or **glucose**, in the blood.
People develop an illness called diabetes if their body does not make enough insulin.

intellect *noun*
The intellect is the thinking power of the **mind**. Having an intellect helps people to reason, understand and remember. Teaching the intellect builds up **intelligence**.
He used intellect, not strength, to solve the problem.

intelligence *noun*
Intelligence describes the skills and knowledge of the **intellect**. It can be measured in a test called an IQ, which stands for Intelligence Quotient. IQ tests ask a person to solve problems and answer questions. The person's scores are compared with the scores of other people of the same age.
Intelligence is mostly inherited, but may be affected by surroundings.

intensive care *noun*
Intensive care is extra health care for very ill people in **hospital**. An intensive care unit in a hospital has special equipment and more doctors and nurses than other hospital departments. The equipment includes a life-support system. This keeps people alive if they are in a **coma** and unable to breathe on their own.
He was in intensive care until he was strong enough to go back to a hospital room.

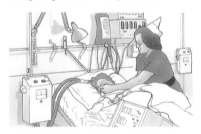

intestine *noun*
The intestine is a long tube that lies coiled in the **abdomen**. It is also called the gut or the bowel. Food from the **stomach** passes through the intestine where it is broken down, or **digested**. The **nutrients** from the food are absorbed into the body through the walls of the intestine, until only the waste matter, or **feces**, is left behind. The intestine is divided into two parts called the **small intestine** and the **large intestine**.
The intestine is about eight metres long.

involuntary system   *noun*

The involuntary system describes the muscles that work on their own in the body. Involuntary muscles move without a person having to think about it. The muscles that control **blood vessels**, the **intestine** and the **stomach** are all involuntary muscles. The **heart** is part of the involuntary system. It beats through every minute of a person's life, even when they are asleep.
People cannot control the muscles of the involuntary system.

iris   *noun*

An iris is part of an **eye**. It is the ring around the **pupil** which gives the eye its colour. It lies in front of the **lens** and behind the **cornea**. Muscles in the iris control how much light passes through the pupil. In dim light, the iris becomes smaller, or contracts. This makes the pupil larger and able to take in as much light as it can. In bright light, the iris grows bigger. This covers up the pupil and makes it smaller. Less light is let into the eye.
It is the iris, not the pupil, that moves.

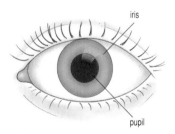

iron   *noun*

Iron is an element. It is a metal which the body needs to stay healthy. Iron is found in foods such as eggs, lentils and lean meat, especially liver. The body uses iron in a part of the blood called **hemoglobin**. If people do not have enough iron in their blood, they develop an illness called **anemia**.
Pregnant women need extra iron to stay healthy.

ischium (plural **ischia**)   *noun*

The ischia are two bones in the **pelvis**. There is one ischium on each lower side of the hipbone. The ischia join together to form an empty circle of bone with two loops at the bottom. The lower ends of these loops are used for sitting. The ischia are joined, or fused, to two other parts of the hipbone called the **pubis** and the **ilium**.
His ischium was fractured in the road accident.

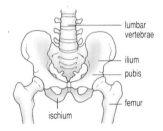

itch   *noun*

An itch is a tickly feeling on the **skin**. Itches make people want to scratch them. Strong itches burn and tingle and feel uncomfortable. Scratching these kinds of itch may break the skin and make it sore. There are many reasons for itching. It can be a **symptom** of dry skin, an **infection** or an **allergy**. Nearly all scabs and spots tend to itch, for example a **chickenpox** rash.
She could not reach the itch in the middle of her back.

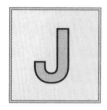

jaundice *noun*

Jaundice is a sign that there is something wrong in the **blood** or the **liver**. The skin and whites of the eyes turn yellow in a person with jaundice. This can be a sign that the **liver** or **gall bladder** is not working properly. It may also be a **symptom** of other kinds of illness. New-born babies often have mild jaundice for a few days. This is harmless and soon disappears.
As he recovered from jaundice, the yellow colour of his eyeballs faded.

jaw *noun*

A jaw is one of two parts of a **face**. The lower jaw runs from beneath each ear to the chin. It forms the lower part of the face and mouth. The upper jaw forms the roof of the mouth and the holes which open out to the nose. It runs up to the bottom edge of the eye sockets and is joined each side to the cheekbones of the skull. Both jaws are made of bone and are called the **jawbones**.
She clenched her jaw so hard her teeth ached.

jawbone *noun*

A jawbone is two parts of a **face**. The **teeth** are set in the jawbone. The upper jawbone is made of two bones joined, or fused, together at the front. They are called **maxillae.** The upper jawbone is attached to the rest of the skull and does not move. The lower jawbone is made of two bones fused together at the front to form the chin. These are called **mandibles**. The upper and lower jawbone are joined together by a **hinge joint** under each ear. Strong muscles in the cheeks move the lower jawbone.
The ball hit him so hard, he was afraid it had broken his jawbone.

jejunum *noun*

A jejunum is a tube inside the body. It is the middle part of the **small intestine**. The jejunum is about 1.2 metres long and joins the other two parts of the small intestine, the **duodenum** and the **ileum**. Juices from the walls of the jejunum help to break down, or **digest**, food before it is passed on through the body.
The jejunum is important for digestion.

joint ► page 84

jugular vein *noun*

A jugular vein is a **blood vessel** in the neck and head. Four main jugular veins carry blood from the head back to the heart.
As he lifted the piano, she saw his jugular vein bulge.

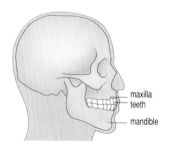

maxilla
teeth
mandible

joint *noun*

A joint is the place where two or more
bones meet. Joints are held in place by
strong, stretchy strips of **connective tissue**.
These are called **tendons** and **ligaments**.
If a joint twists too far, the ligaments may
tear, causing a sprain. A joint that is pulled
out of place is **dislocated**.
Her joints became stiff with arthritis.

A smooth layer of cartilage covers the ends of bones that
move over each other. A liquid called synovial fluid helps
the joints to move smoothly.

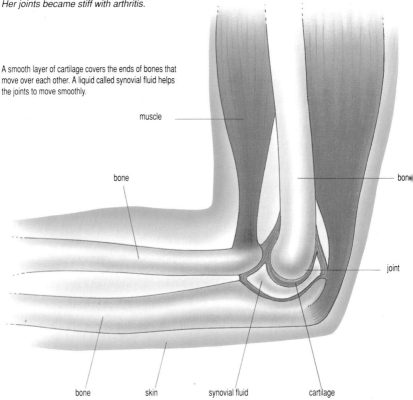

muscle

bone

bone

bone

joint

bone   skin   synovial fluid   cartilage

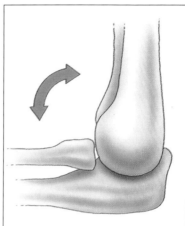

Ball and socket joints can move in a circle. Shoulders and hips have ball and socket joints.

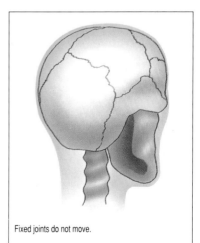

Hinge joints move like a door hinge. Knees and fingers have hinge joints.

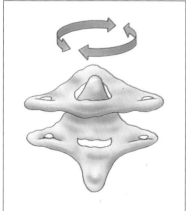

Pivot joints allow a rotating movement. The elbow and the first two vertebrae in the spine form pivot joints.

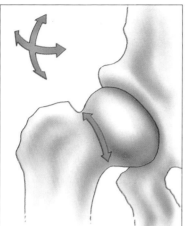

Fixed joints do not move.

keratin *noun*
Keratin is a tough substance which is made
of **protein**. It is found in the outer layer of the
skin and makes it waterproof. Keratin builds
up where the skin wears away the most,
such as the hands and feet. Nails, teeth and
hair also contain keratin.
*There are no nerves or blood vessels in
keratin.*

kidney ► page 87

kidney machine ► **dialysis**

knee *noun*
A knee is part of a **leg**. It is the **joint** that
joins the upper leg to the lower leg. Knees
work like a hinge and allow the legs to bend
in half. The front of each knee is covered by
a small plate of bone called the **kneecap** or
patella. Strong ligaments either side of the
knee and inside the joint help to keep it from
becoming **dislocated**.
*The knee has to support the weight of the
whole body.*

knee-jerk test ► **reflex**

kneecap *noun*
A kneecap is a small, disc-shaped bone in
front of the knee. It is also called the patella.
It is in the middle of the **tendon** that joins the
thighbone, or **femur**, to the shinbone, or
tibia. The inside of the kneecap is covered
with **cartilage**, and forms part of the **joint**
which allows the leg to bend.
*The cartilage of his kneecap was damaged,
which made walking painful.*

Kwashiorkor *noun*
Kwashiorkor is an illness found in children.
It is caused by not having enough food to
eat. A child develops a swollen stomach and
feet. The skin becomes dry, pale and scaly.
Children with Kwashiorkor cannot grow
properly. Without treatment their body
becomes very thin and the skin shrinks and
wrinkles over the bones.
*Many children suffer from Kwashiorkor
during a famine.*

kidney   *noun*

A kidney is one of two **organs** in the body.
The kidneys are about 10 centimetres long
and lie each side of the **spine** at the back of
the abdomen. They filter the **blood** and keep
it clean. The kidneys also collect water the
body does not need. The water and waste
material filtered by the kidneys form a yellow
liquid called **urine**.
A person can live with only one kidney.

A filtering unit

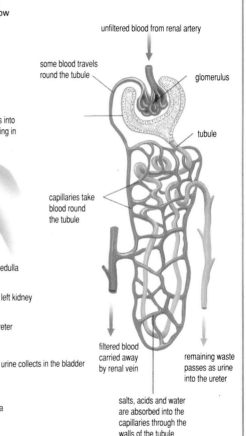

unfiltered blood from renal artery

some blood travels
round the tubule

glomerulus

each artery branches into
smaller arteries, ending in
a filtering unit

blood flows into the kidneys
through the renal artery

tubule

capillaries take
blood round
the tubule

adrenal
gland

medulla

right kidney

left kidney

artery

ureter

vein

urine collects in the bladder

filtered blood
carried away
by renal vein

remaining waste
passes as urine
into the ureter

urethra

salts, acids and water
are absorbed into the
capillaries through the
walls of the tubule

labia   *plural noun*

The labia are part of the outer sexual organs, or **vulva**, of a female. The labia are long folds of skin arranged in two pairs, one inside the other. The outer labia fold partly over the inner labia and surround the opening of the **vagina**. The inner labia help to protect the **clitoris** and the opening of the **urethra**.
During puberty, the outer labia grow a covering of pubic hair.

lacrimal gland   *noun*

A lacrimal gland is one of the two **glands**, one above each eye, that produce tears.
The muscles round the lacrimal glands tightened, and his eye filled with tears.

lactose   *noun*

Lactose is a kind of sugar found in milk. Some people are unable to **digest** the lactose in milk. This gives them stomach pain and **diarrhea.** People who react in this way to lactose should not drink milk or eat food made with milk, such as butter, cheese and cream.
She cannot eat dairy products because she cannot digest lactose.

laparoscopy   *noun*

A laparoscopy is a look inside the **abdomen**. This is done by a doctor with a lighted tube called a laparoscope. The tube is slipped into a small cut in the skin beside the **navel**. A doctor can look at the liver, intestine, bladder, uterus and ovaries by carrying out a laparoscopy.
During the laparoscopy, the doctor found a problem with the man's liver.

large intestine   *noun*

The large intestine is joined to the **small intestine** and runs down to the opening of the **anus**. It is about 1.5 metres long. The three sections of the large intestine are the **cecum**, **colon** and **rectum**. Water is taken in, or **absorbed**, from digested food in the large intestine. The waste left behind is passed out through the anus as **feces**.
Colitis is a disease of the large intestine.

larynx   *noun*

A larynx is a part of the **throat**. The larynx reaches from the root of the tongue, at the back of the mouth, down to the opening of the **windpipe**, the airway to the **lungs**. The **vocal cords** inside the larynx help people to make sounds.
At puberty, his larynx grew bigger .

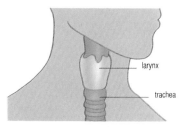

leg   *noun*

A leg is the lower **limb** of the body. The two legs are joined to the **pelvis** by the hip **joints**, and end in the **feet**. Strong **muscles** attached to the leg bones help them move.
The legs bend at the hips and knees.

lens   *noun*

A lens is part of an **eye**. It is the soft, see-through disc that lies behind the **cornea** and the **iris**. Light is focused through the lens before it passes to the retina at the back of the eye. Muscles change the shape of the lens to focus at a long or a short distance.
A cataract is a disease that causes the lens to become cloudy.

leprosy *noun*
Leprosy is a **disease** found in tropical areas.
It develops slowly, attacking the skin and
nerve endings. Patches of skin change
colour, become thicker and lose all sense of
feeling. In a more serious form of leprosy, the
skin develops lumps and open sores.
Leprosy is caused by **bacteria**. It is
infectious, but there is not much danger of
catching it from another person.
Leprosy can be treated with drugs.

leukemia *noun*
Leukemia is a **disease** of the **white blood
cells**. It is a kind of **cancer** that starts in the
bone marrow where blood is made. White
blood cells are part of the body's defence, or
immune system. Some kinds of leukemia
develop suddenly, but others develop slowly
over a number of years. Symptoms of
leukemia include weakness, **anemia**,
bruising under the skin and **nosebleeds**.
*The girl with leukemia had a bone marrow
transplant.*

leukocyte ► white blood cell

lice (singular **louse**) *plural noun*
Lice are tiny insects that live on mammals.
Head lice are easy to catch from another
person. They climb to the next head and lay
their eggs on hairs. Head lice make the head
itch. Medicated shampoos and lotions get rid
of head lice quite easily. The tiny white eggs
should be combed out of the hair with a
fine-toothed comb and the treatment
repeated after a week.
*You can catch head lice even if your hair is
very clean.*

life cycle *noun*
A life cycle describes the stages of life from
birth to **death**. Growing from a **baby** to a
child and then to an **adolescent** are the
three first stages. An adolescent grows to an
adult. During adulthood, people reproduce,
and more babies are born. This starts a new
life cycle. In this way, a life cycle is the never-
ending circle of birth, death and birth which
keeps the human race alive.
*A human body goes through many changes
during its life cycle.*

life expectancy *noun*
Life expectancy is the length of time a person
may expect to live. It is also called a lifespan.
A person's life expectancy depends on how
long their parents lived, and how well their
health is looked after. People who live in an
environment with plenty of clean air, water,
food and good medical care have a long life
expectancy. Females usually have a higher
life expectancy than males.
*The average life expectancy in Africa is
shorter than in Europe.*

lifestyle *noun*
A lifestyle is the way a person lives their life.
A healthy lifestyle should include regular
exercise, a balanced **diet** of food and rest
and relaxation.
*He changed to a healthier lifestyle when the
doctor told him he had high blood pressure,
or hypertension.*

life-support system ► intensive care

ligament *noun*
A ligament is a strong, fibrous band of
tissue. Ligaments hold the ends of **bones** or
joints together. They also hold and support
other organs inside the body. Ligaments are
not elastic, but they are stretchy enough to
allow joints to move. A **sprain** happens when
a ligament is twisted or torn. Ligaments are
often slow to heal.
*During the football match, he tore a knee
ligament and had to limp off the pitch.*

limb   *noun*
A limb is an **arm** or a **leg**. Arms are called
upper limbs and legs are called lower limbs.
After the crash, her lower limbs were put in
plaster so they could mend properly.

lip   *noun*
A lip is a fleshy edge or fold of skin around
an opening in the body. The **mouth** is
surrounded by a pair of lips which open and
close. The two pairs of lips that surround the
vagina and urethra are called the **labia**
He put his lips to the glass and drank all the
water.

liver ► page 91

lobe   *noun*
1. A lobe is a section of an **organ**. Lobes are
usually rounded areas that are separated by
bands of tissue. The brain, liver and lungs
are all divided into lobes.
The doctor felt the lobe of his liver under the
rib cage.
2. The lobe of the ear is the soft, lower part
of the outer ear.
A large earring hung from the lobe of her ear.

long-sightedness   *noun*
Long-sightedness, or far-sightedness, is
being able to see things clearly far away, but
not close up. It is caused by light being
focused behind the **retina** at the back of the
eye instead of on it. Long-sightedness
sometimes runs in families, but it may also
develop in anyone over the age of about 40.
Wearing spectacles or contact lenses helps
long-sightedness.

lumbago   *noun*
Lumbago is a kind of **backache**. It is felt in
the muscles around the lower parts of the
spine, especially when lifting something or
bending over. Lumbago can be caused by a
strain which heals with rest. It may also be a
symptom of another illness or back problem.
Many old people have difficulty bending over
because of lumbago.

lumbar   *adjective*
Lumbar describes anything to do with the
lower **back**. The five lumbar vertebrae are in
the lower back, between the bottom **rib** and
the **pelvis**. A lumbar puncture is a way of
drawing out fluid from this part of the **spine**
with a needle.
The car seat had a lumbar support which
made it more comfortable to sit in.

lung   *noun*
A lung is one of the pair of **organs** used for
respiration. The lungs lie either side of the
heart in the **cavity** of the chest, or thorax. Air
is breathed in through the mouth and travels
down the **trachea**, **bronchi** and
bronchioles. At the end of the bronchioles,
the air collects in millions of tiny air sacs
called **alveoli**. The lungs are surrounded by
two layers of thin membrane called **pleura**.
The boy almost drowned by breathing water
into his lungs.

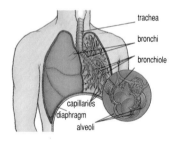

trachea
bronchi
bronchiole
capillaries
diaphragm
alveoli

lymphatic system   *noun*
The lymphatic system is a network of small
vessels that contain a clear, watery fluid
called lymph. Lymphatic vessels collect fluid
that has seeped from the **capillaries**, and
return it to the **bloodstream**. The lymphatic
system is also part of the body's **immune**
system. Small lumps called lymph nodes
produce some of the **white blood cells** and
antibodies that fight infection.
Bacteria and waste are filtered from the
lymph as it passes through the lymph nodes.

liver *noun*

The liver is the largest **organ** in the body. It is a **gland** which lies on the top right-hand side of the **abdomen**, under the ribs. The liver helps to digest food and keep the blood clean. **Bile** is made in the liver and passed to the **intestine** to help break down food. The liver also takes in and re-uses **nutrients** and other substances from the blood. Poisons in the blood, such as alcohol, are destroyed by the liver.
Most digested food goes to the liver before it goes to the rest of the body.

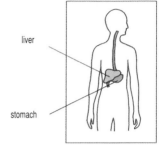

liver

stomach

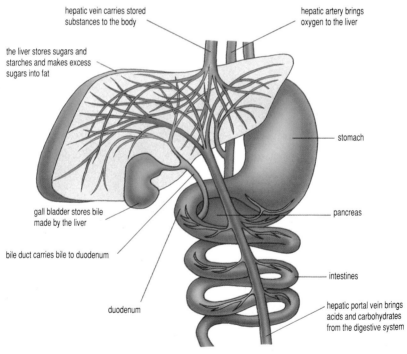

hepatic vein carries stored substances to the body

hepatic artery brings oxygen to the liver

the liver stores sugars and starches and makes excess sugars into fat

stomach

gall bladder stores bile made by the liver

pancreas

bile duct carries bile to duodenum

intestines

duodenum

hepatic portal vein brings acids and carbohydrates from the digestive system

malaria   *noun*
Malaria is a **disease** which attacks **red blood cells**. It is caught from the bite of mosquitoes that carry the disease, usually in a tropical country. The first sign of malaria is a bout of shivering. This is followed by several hours of high **fever**, sweating, vomiting and headache. The symptoms tend to come back every few days until the illness is treated.
He took tablets to avoid catching malaria.

male   *adjective*
Male describes anything to do with boys and men. A male child is a boy. Male is the opposite of a **female**, which means girl or woman.
After having three girls, the parents longed for a male child.

malignant   *adjective*
Malignant describes a dangerous illness that may become worse. **Cancer** is a malignant growth, or tumour, in the body. Malignant tumours can sometimes be removed surgically. The opposite of malignant is **benign**.
Someone with a malignant disease needs sympathetic care.

malnutrition   *noun*
Malnutrition is ill health caused by not eating enough food. It may also be caused by eating too little of the right kinds of food. Children with severe malnutrition may develop an illness called **Kwashiorkor**.
Malnutrition is found in poorer countries where food is scarce.

mammary gland   *noun*
A mammary gland is another term for a **breast**. In women, the two mammary glands grow and develop on the front of the chest during **puberty**. Each fully-grown breast contains 15 to 20 milk glands surrounded by fatty tissue. Tubes from the milk glands run towards the tip of the breast, or **nipple**. The size and shape of mammary glands vary.
After a woman has given birth, the mammary glands start making milk to feed the baby.

mandible ► **jawbone**

marrow ► **bone marrow**

massage   *noun*
Massage is a way of rubbing and stroking the skin. This can be done by another person's hands, a machine, or flowing water. Massage helps to ease tense and painful muscles underneath the skin. It also improves the circulation of blood and makes a person feel calm and relaxed. Massage is sometimes used to help rebuild muscles after a long illness.
It is good to relax with a massage after a hard day.

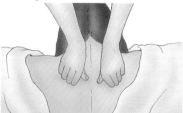

mature   *adjective*
Mature describes being fully grown. It can describe parts or all of the body. It can also describe personality and behaviour. The opposite of mature is **immature**.
The teacher told the boy to behave in a more mature way.

maxilla ► **jawbone**

measles *noun*
Measles is a common childhood **disease** which is caused by a **virus**. The first symptoms are a fever, sore throat, cough and runny nose. After four days, blotchy pink spots appear over the face and neck, and spread over the rest of the body. Measles is catching, or contagious, from four days before until five days after the rash appears. There is a vaccine which protects children from getting measles.
Every child should be vaccinated against measles.

medicine *noun*
Medicine is the study and treatment of disease. It is also a **drug** or any other substance used to treat illness. **Doctors** study and practise medicine. They also prescribe medicine for patients.
Many kinds of medicine are flavoured to make them more pleasant to swallow.

melanin *noun*
Melanin is a dark colour, or pigment. It is found in the skin, hair and the iris of the eye. Melanin protects the body from the sun by turning the skin darker.
As the girl sunbathed, melanin turned her skin a shade darker.

melanoma *noun*
Melanoma is a cancer of the **skin**. A harmful growth, or **malignant tumour**, forms in a pigment cell of the skin or eyes. The growth may spread to the lymph glands and to the rest of the body and can be fatal. Moles and freckles that changes colour or shape, itch or bleed need to be seen by a doctor.
People with fair skin who spend long in the sun are at risk of developing melanoma.

membrane *noun*
A membrane is a skin inside the body. Membranes are made of thin layers of tissue and used to cover and separate **organs**. They also line the inside of tubes and **cavities**, inside the body.
Fluid is filtered through a membrane in the kidney.

memory *noun*
Memory is a store of information in the **brain**. It is also the ability to remember the information in the right order. Some information is stored in the memory for a short time. Other things remain there for a long time. The part of the brain which controls memory is called the **cerebral cortex**. Memory is part of the thinking power, or **intellect**, of the mind.
She recited the poem from memory.

meningitis *noun*
Meningitis is a disease caused by a **bacteria** or a **virus**. It is an **inflammation** of the **membranes** that cover the brain and spinal cord. The first signs may be a sore throat and slight symptoms of influenza. This is followed by a severe headache, fever and stiff neck. The eyes may be very sensitive to light. Vomiting, convulsions and coma may follow in serious cases.
The child was rushed to the isolation ward of the hospital as meningitis was suspected.

menstrual cycle *noun*
The menstrual cycle begins when a girl reaches **puberty**. It is controlled by **hormones**. Every month an egg from one of the two **ovaries** travels along a **Fallopian tube** towards the **uterus**. This is called **ovulation**. A fertilized egg will settle into the lining of the uterus to grow into a baby. If the egg is not fertilized, the egg and lining are passed out through the **vagina** as blood. The menstrual cycle is counted from the first day of one period of bleeding to the first day of the next.
The menstrual cycle lasts about 28 days.

menstruation *noun*
Menstruation is a part of the **menstrual
cycle** of a female. It is the period of bleeding
from the **vagina** when the lining of the uterus
is shed. This usually lasts from three to
seven days. The flow of blood is normally
heavier in the first two days.
*She started her menstruation at the age of
eleven.*

mental illness *noun*
Mental illness refers to disorders of the
mind. Doctors who treat people with mental
illnesses are called **psychiatrists**. There are
many causes for mental illness. Some kinds
run in families and are passed through
genes. Others are due to problems during a
person's life. Damage to the brain can also
cause mental illness. Some mental illnesses
only last a short time and can be completely
cured.
*Mental illness affects behaviour and the way
a person thinks.*

metabolic rate *noun*
A metabolic rate is the speed at which a
body burns up **energy**. Every part of the
body is using up energy all the time, even
when a person is asleep. People get their
energy from food and from fat stored in the
body. A person's metabolic rate is controlled
by **hormones** produced in the **thyroid
gland**.
*Some people burn up energy faster than
others and have a higher metabolic rate.*

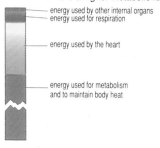

energy used by other internal organs
energy used for respiration

energy used by the heart

energy used for metabolism
and to maintain body heat

metabolism *noun*
Metabolism is all the work carried out by the
body. The body grows, heals, repairs and
replaces tissue. Chemicals and other
substances are produced by the body. At the
same time, the body is breaking down food,
to provide energy to do all these things. This
whole process is called metabolism.
*Some people's metabolism works slowly and
can result in their being overweight.*

metacarpals ► **hand**

metatarsals ► **foot**

microbiology *noun*
Microbiology is the study of tiny living things
that can only be seen under a microscope.
These include **bacteria** and **viruses**. Medical
microbiologists study diseases caused by
bacteria and viruses. They also research into
ways to prevent and cure these diseases.
Dental microbiologists look at the germs that
are found in the mouth and on the teeth.
*He worked on medical research after
studying microbiology at college.*

microsurgery *noun*
Microsurgery is a kind of surgical operation.
Microscopes are used so a surgeon can
operate on delicate parts of the body, such
as nerve endings, or the eye. The surgeon
uses specially adapted instruments in
microsurgery. Sometimes, the operation is
displayed on a television screen.
*The doctor used microsurgery to fix on the
boy's severed finger.*

middle ear *noun*
A middle ear is part of an **ear**. The **eardrum**
separates the middle ear from the outer ear.
The middle ear is filled with air from the back
of the nose by the **Eustachian tube**. Sound
waves pass through the eardrum to three tiny
bones in the middle ear called the **ossicles**.
The ossicles vibrate with the sound waves
and pass them on to the inner ear.
Fluid in the middle ear can cause deafness.

midwife   *noun*

A midwife is a person trained to help women in **childbirth**. Midwives also look after women during pregnancy when they are expecting a baby. They check how the baby is growing and the health of the mother. After the baby is born, they care for the mother and baby for the first few days.
The midwife rushed to the house to help the woman give birth.

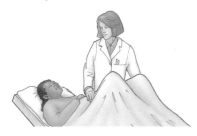

migraine   *noun*

A migraine is a type of **headache**. It is a severe, throbbing kind of **pain** which lasts hours or days. It is often in only one side of the head. A person with migraine may feel sick or vomit and the light may hurt their eyes. Before a migraine starts, some people have a strange taste or smell. Migraines may be caused by a reaction to certain foods, bright lights or loud noises.
A person with a migraine should lie down in a dark room.

milk   *noun*

Milk is a liquid food. It is made by the **mammary glands** of a female mammal after the birth of a baby. Milk is a complete food, which means that it has most of the **nutrients** people need to live. Human milk is the best kind of food for babies for the first few months. Cow's milk is an important part of most people's diets after the age of one year.
The chef made the custard with milk, eggs and sugar.

milk teeth ▶ tooth

mind   *noun*

Mind is another word for the intellect of the **brain**. It is the part of a person which thinks, has feelings and understands. The mind is the part of a person that is not the body. All the parts of the mind make up a person's personality, or what kind of a person they are. The body and the mind together make up a whole person.
She had a vivid picture of the accident in her mind.

mineral   *noun*

A mineral is one of many substances found in soil, water, rock and metal. Some minerals are needed to keep the body healthy. **Calcium** is important for teeth and bones, and **iron** is needed for healthy blood. A balanced **diet** provides all the minerals a person's body needs. It is not usually necessary to take extra minerals in tablet form.
You need to eat foods containing the mineral calcium to develop strong bones.

miscarriage   *noun*

A miscarriage is the loss of a baby in the early part of **pregnancy**. Most miscarriages happen because there is something wrong with the baby. The first signs of a miscarriage are pains in the **abdomen** and bleeding from the **vagina**.
A year after her miscarriage, the woman gave birth to a healthy baby.

molar ▶ tooth

mole   *noun*

A mole is a dark spot or place on the **skin**. Moles vary in size, shape and colour. They can be flat and smooth or raised and hairy. Moles are caused by a dark pigment in the skin called **melanin**. Most moles develop as a person is growing into an adult.
The doctor removed the large mole growing on his chin.

movement *noun*

Movement is the act of changing from one position to another. The body is able to make many different kinds of movement through a network of **muscles**, **tendons** and **ligaments**. Every movement a person makes is a result of messages from the **nerves** to the muscles. Movements are controlled by the **brain**, and the **ears** and **eyes** are also often involved.
With a jerky movement, he swept the cup off the table.

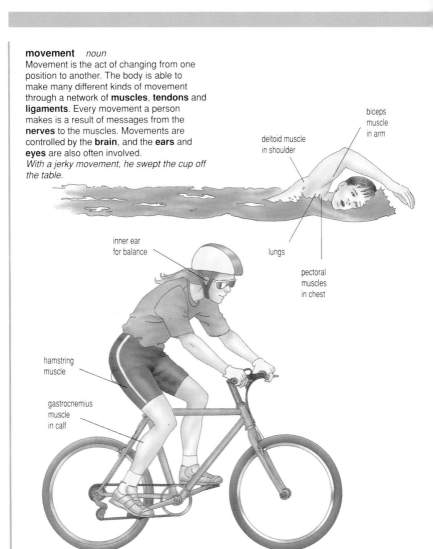

biceps muscle in arm

deltoid muscle in shoulder

inner ear for balance

lungs

pectoral muscles in chest

hamstring muscle

gastrocnemius muscle in calf

inner ear for sound and balance

thigh muscles

pelvis

gastrocnemius muscles in calves

eyes co-ordinate movements of hand

small muscles in hand

eyes co-ordinate movements of hands

inner ear for sound

finger muscles

muscles in forearm

muscle *noun*

Muscle is strong, stretchy **tissue**. There are over 600 muscles arranged around the frame of the **skeleton** and inside the body. Muscles are able to contract, or tighten, into a shorter, fatter shape. By doing this they make parts of the body move. The muscles attached to bones usually work in pairs, such as the **biceps** and **triceps**.
Muscle makes up nearly half the body weight of an adult.

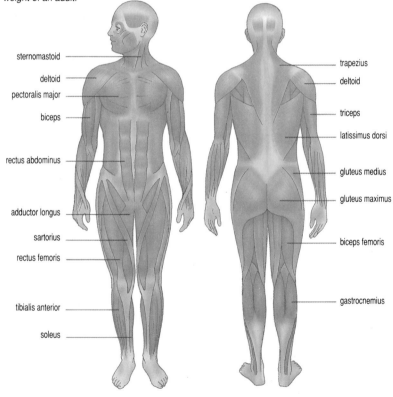

sternomastoid
deltoid
pectoralis major
biceps
rectus abdominus
adductor longus
sartorius
rectus femoris
tibialis anterior
soleus

trapezius
deltoid
triceps
latissimus dorsi
gluteus medius
gluteus maximus
biceps femoris
gastrocnemius

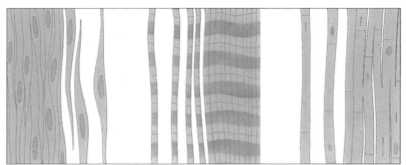

Involuntary or smooth muscle in the stomach and intestines. These muscles work automatically. They cannot be controlled.

Voluntary, or striated (striped) muscles, for moving the skeleton. These muscles can be controlled at will.

Cardiac muscles in the heart. These muscles work automatically. They cannot be controlled.

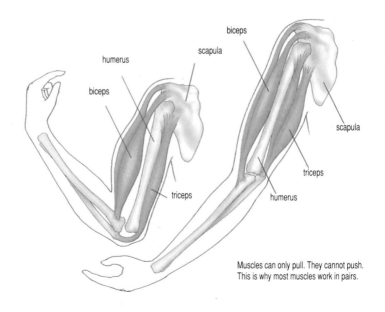

biceps

scapula

humerus

biceps

scapula

triceps

triceps

humerus

Muscles can only pull. They cannot push. This is why most muscles work in pairs.

molecule *noun*

A molecule is a small particle. Molecules are the smallest complete part of any substance, and contain two or more atoms that are joined together.

The molecules in a liquid or a gas move about easily.

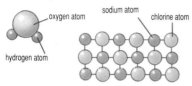

water molecule   salt molecule

motor *noun*

Motor describes something to do with movement in the body. Motor nerves send signals to muscles to make them move.

The disabled boy needed help strengthening the motor control in his legs.

motor cortex *noun*

A motor cortex is part of a **brain**. It is part of the outer layer of the largest part of the brain. The motor cortex controls the movements a person wants to make. It does this by sending messages along motor **nerves** to the muscles that move the body.

His arm was paralysed because his motor cortex had been damaged.

motor reflex ► reflex

mouth *noun*

A mouth is part of a **face**. It is the opening in the body formed by the upper and lower **jawbone**. **Lips** surround the entrance and open and close the mouth. Inside the mouth are the **gums**, **teeth**, **palate** and **tongue**. The soft skin inside the mouth is kept moist with **saliva** and **mucus**. Mouths are used for eating and speaking and sometimes also for breathing.

Deaf people can read by the movements made by the lips and mouth.

movement ► page 96

mucous membrane *noun*

A mucous membrane is a moist layer of **skin**. The surface is kept moist by tiny glands that release a sticky fluid called **mucus**. Mucous membranes line many parts of the inside of the body. They also line the inside of the openings of the body, such as the **mouth** and the **vagina**.

The hot food burned the mucous membrane in her mouth.

mucus *noun*

Mucus is a sticky fluid made by a **mucous membrane**. Mucus protects and cleans the surface of a membrane. It also makes it slippery and smooth. Some mucus contains **enzymes** which help to break down substances, such as **food**.

She coughed up mucus when she had bronchitis.

multiple sclerosis *noun*

Multiple sclerosis is a **disease**. It attacks the **nerves** in the **brain** and **spinal cord**. The development of the disease can be very slow. Symptoms start with tingling and numb feelings in the skin. A person may then lose control of their **muscles**. People recover between attacks, but each time the illness comes back, the symptoms are worse and stay longer.

There is no cure for multiple sclerosis.

mumps *noun*

Mumps is a common childhood **disease** caused by a **virus**. It can be a more serious illness when caught by an adult. The first sign of mumps may be a **headache**, **fever** and vomiting. A day or two later, the **glands** around one or both ears become swollen and painful. It may be painful to open the mouth and swallow. The swelling can last for five or six days. Mumps remains **infectious** for about three days after the swelling goes down.

There is a vaccine to protect children from catching mumps.

muscle ► page 98

muscular *adjective*
Muscular describes anything to do with
muscles. Regular **exercise** develops strong,
muscular bodies.
*The champion weight lifter had very
muscular thighs.*

muscular dystrophy *noun*
Muscular dystrophy is a group of **diseases**
that affect the **muscles**. The muscles
attached to the **bones** waste away and do
not work properly. The disease gradually
gets worse over a number of years. These
kinds of muscle disease are hereditary, or
passed on in families through faulty **genes**.
More boys than girls are born with this kind
of disease. There is no cure for muscular
dystrophy.
*People with muscular dystrophy feel weak
and find it difficult to move.*

myopia *noun*
Myopia is being able to see objects clearly
when they are near, but not at a distance.
The non-medical name for myopia is
short-sightedness. People suffer from
myopia if the lens of the eye is too curved, or
the eyeball too long. Light is then focused in
front of the **retina** at the back of the eye
instead of on it.
*Wearing spectacles or contact lenses
corrects myopia.*
myopic *adjective*

nail *noun*
A nail is a piece of hard **tissue** made of
keratin. Nails grow at the end of each **finger**
and **toe**. The whitish, semi-circular area near
the base is called the lunula. Toenails are
tougher than fingernails and grow more
slowly. A change in the colour or shape of
nails can be a sign of an illness.
*She filed her nails regularly to keep them
short.*

narcotic *noun*
A narcotic is a strong **drug** which relieves
pain and makes a person feel drowsy.
Narcotic drugs can be dangerous when
taken without the advice of a **doctor**.
A narcotic drug can become addictive.
narcotic *adjective*

nasal *adjective*
Nasal describes anything to do with the
nose. A nasal sound can be made by closing
the back of the throat with the **tongue** and
humming. Air passes through the nasal
passages in the nostril when a person
breathes.
She talked with a nasal sound.

nausea *noun*
Nausea is the feeling of wanting to be sick,
or **vomit**. Bringing up the contents of the
stomach often gives relief. Nausea may be a
symptom of stomach upsets, **pregnancy**,
travel sickness or shock. It can also be a
symptom of other kinds of illness.
*The man felt nausea when the sea was
rough.*
nauseous *adjective*

nervous system *noun*

The nervous system is a network of **nerve cells** called neurons, that send information to and from all parts of the body. The **brain**, **brain stem** and **spinal cord** are the central parts of the nervous system. Messages, or nervous impulses, travel along nerves that connect the brain to the rest of the body. All the activities of the body, such as breathing and moving, are controlled by the nervous system.

Motor neuron disease affects the nervous system.

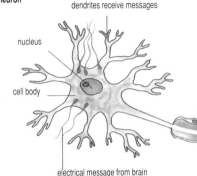

a neuron

dendrites receive messages

nucleus

cell body

electrical message from brain

brain

central nervous system

spinal cord

brain stem

peripheral nerves

sympathetic nerves

The central nervous system consists of the brain and the spinal cord.

The peripheral nervous system involves nerves that are linked with the central nervous system and branch through the body.

The autonomic nervous system is responsible for controlling the automatic movement of internal organs. Part of this system is called the sympathetic system. Sympathetic nerves form two cords that run parallel with the spinal cord. They are responsible for actions such as speeding up the heartbeat.

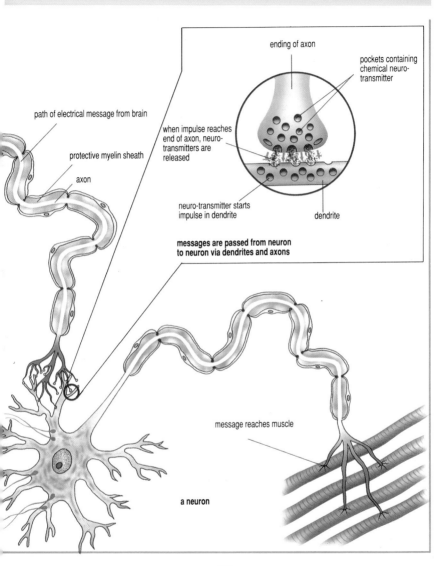

ending of axon

pockets containing chemical neuro-transmitter

path of electrical message from brain

protective myelin sheath

axon

when impulse reaches end of axon, neuro-transmitters are released

neuro-transmitter starts impulse in dendrite

dendrite

messages are passed from neuron to neuron via dendrites and axons

message reaches muscle

a neuron

navel *noun*
A navel is the healed end of the **umbilical cord** in the middle of the outside of the **abdomen**. Navels may be a small hollow in the skin or a small, knotted lump that sticks out. The navel is the point where a person was joined to their mother in the uterus by the umbilical cord. This cord is cut soon after birth. The stump that is left shrivels away, falls off and leaves a navel.
The man was so fat he could not see his navel.

near-sightedness ► myopia

neck *noun*
The neck is the narrow part of the body between the **head** and the **shoulders**. The seven bones of the neck, or **cervical vertebrae**, support the head and allow it to move. **Blood vessels** run through the neck and connect the **brain** and other parts of the head with the rest of the body. The organs in the neck include the **trachea**, **esophagus** and **thyroid gland**.
She had a whiplash injury which caused a stiff neck.

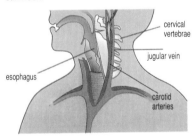
cervical vertebrae
jugular vein
esophagus
carotid arteries

nephron *noun*
A nephron is a tiny tube in a **kidney**. Over a million nephrons in each kidney filter water, salts and glucose from the **blood**. The waste liquid left behind is then passed from the kidney and out of the body as **urine**.
A kidney dialysis machine does the job of the nephrons.

nerve *noun*
A nerve is a bundle of strands in the **nervous system**. Messages pass through **nerves** on their way to and from the **brain**. The brain uses nerves to pass on instructions to every part of the body. The main nerves work in pairs and start from the **brain stem** and the **spinal cord**. These branch out through the body in long strands covered by a sheath of **connective tissue**.
The pinched nerve at the base of his spine caused him great pain.

nerve ending *noun*
A nerve ending is the end of one of the strands of a **nerve**. Nerve endings touch other nerves, **glands**, **muscles** and other **tissues** in the body. **Motor nerve** endings tell muscles to move. **Sensory nerve** endings in the **skin** pick up messages of pain, heat and cold.
The exposed nerve ending caused bad toothache.

nervous system ► page 102

neuron *noun*
A neuron is a special type of **cell**. The **brain**, **spinal cord** and **nerves** are all made from neurons. These kinds of cell can send and receive signals, or nerve impulses, all over the body. Billions of different kinds of neurons make up the **nervous system** in the body.
The message was sent from the brain to the muscle through a string of neurons.

nipple *noun*
A nipple is part of a **breast**. It is the small, brownish or pink teat in the centre of each breast. Nipples are surrounded by a ring of the same coloured skin called the areola. During puberty, the nipples and areolae in a girl become larger and darker. Milk is sucked out of tiny openings in the nipples during **breast-feeding**.
The baby fed from its mother's nipple.

nose ► page 105

nose *noun*

The nose is the central feature of a face. Two small bones meet in the centre to form the bridge of the nose. The tip of the nose and nostrils are shaped with **cartilage** and skin. The inside of the nose is lined with a **mucous membrane** and joins the airways at the back of the thoat. Air is warmed, cleaned and moistened as it travels through the nose on its way to the **lungs**. Smells are picked up by nerve endings at the top of the inside of the nose.

Noses are for breathing and smelling.

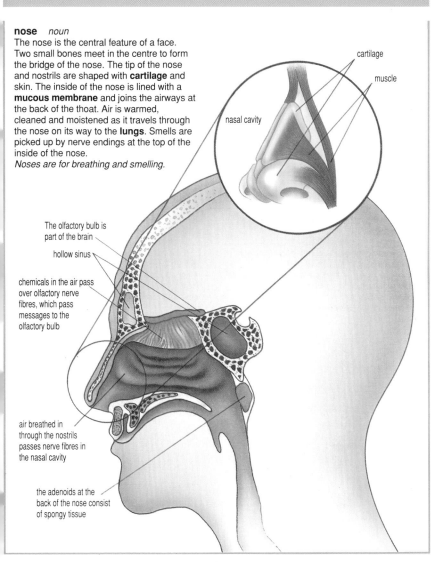

cartilage

muscle

nasal cavity

The olfactory bulb is part of the brain

hollow sinus

chemicals in the air pass over olfactory nerve fibres, which pass messages to the olfactory bulb

air breathed in through the nostrils passes nerve fibres in the nasal cavity

the adenoids at the back of the nose consist of spongy tissue

nosebleed *noun*
A nosebleed is bleeding from one or both **nostrils**. Sneezing, blowing or picking the nose can all make it bleed. Pinching the nose should stop the flow of blood. Nosebleeds may also be a **symptom** of a blood disorder or other illness.
The ball hit him on the nose, and gave him a nosebleed.

nostril *noun*
A nostril is one of the two openings in the **nose**. Nostrils are divided by a piece of thin **bone** and **cartilage**. The two channels of the nostrils lead towards the nasal cavity. This is the space that lies between the floor of the **skull** and the roof of the **mouth**.
Delicious smells from the kitchen reached her nostrils.

nucleus *noun*
A nucleus is the middle part of a **cell**. The nucleus of each cell contains **chromosomes**, **genes**, and the **DNA** that forms them.
The nucleus is the control point of each cell.

nurse *noun*
A nurse is a person trained to treat people who are ill. Some nurses have special training to look after certain kinds of illness or people. For example, geriatric nurses look after old people. Nurses also help carry out operations and check-ups.
She went to medical school to learn to be a nurse.

nutrient *noun*
A nutrient is a substance in food. The body needs nutrients for **energy**, **growth** and good **health**. The important nutrients in food are **proteins**, **carbohydrates**, **fats**, **vitamins** and **minerals**.
The nutrients in processed food are noted on the outside of the packet.

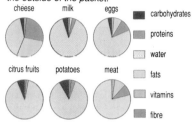

nutrition *noun*
Nutrition is the study of **food** and **health**. It is also the whole process of how people prepare, eat and **digest** food to keep the body in good working order. People need to eat a balanced **diet** to stay healthy. This must contain the right amounts of **nutrients**, **fibre** and fluids.
The amount of nutrition in vegetables is reduced by over-cooking.

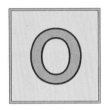

olfactory   *adjective*
Olfactory describes anything to do with the **nose** and the sense of smell.
His olfactory senses smelled the dinner.

operation   *noun*
An operation is a kind of medical treatment. Operations repair, replace or remove **organs** inside the body. People are given **drugs** called **anesthetics** to deaden pain or make them **unconscious**. A surgeon cuts open the skin and after the operation, sews up any cut in the skin to help it to **heal**.
Her eye operation made her see clearly again.

optic   *adjective*
Optic describes anything to do with the **eye** or **sight.**
Visual messages are transmitted to the brain by the optic nerves.

optician   *noun*
An optician is a person who helps to correct **eye** problems. Opticians make and fit lenses for spectacles, and also **contact lenses**.
The optician made sure her lenses were comfortable.

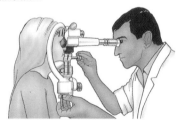

oral   *adjective*
Oral describes anything to do with the mouth. Medicine taken orally is put into the mouth and swallowed.
Brushing the teeth is an important part of oral hygiene.

orbit   *noun*
An orbit is one of two bony sockets in the face. The **eyeballs** fit into the orbits in a bed of fatty **tissue**. Six **muscles** attach the eyeball to the orbit and allow it to move around. The eyes are linked to the **brain** through a hole at the back of each orbit.
He felt a pain as he moved his eyes in their orbits.

organ   *noun*
An organ is any one complete part of a **body**. Organs, such as the **liver** or the **heart**, carry out certain jobs in the body. The **eyes** are organs of **sight** and the **ears** are organs of **hearing** and **balance**. Organs may be made of any kind of two or more **tissues** and be any size or shape.
The heart and the brain are two of the most important organs in the human body.

organic   *adjective*
1. Organic describes anything to do with an organ.
Pneumonia is an organic disease.
2. Organic also refers to natural substances formed in or from plants or animals.
Organic gardeners do not use chemical fertilizers.

ossicles   *plural noun*
Ossicles are three tiny bones in the **middle ear**. They vibrate when sound waves pass through the **eardrum**. Two small muscles attach the ossicles to the bone around them. When a loud sound passes through the middle ear, the muscles tighten. This stops the ossicles from vibrating too much and damaging the inner ear.
The ossicles are called the hammer, the anvil and the stirrup.

ossification *noun*
Ossification is the forming of **bone**. This happens as a baby develops in the **uterus** before it is born. Ossification carries on as children grow into adults. During **adolescence**, bones reach their full size. A person grows taller and broader in this last stage of ossification.
The bone in a broken leg knits together by ossification.

osteopathy *noun*
Osteopathy is a way of treating **illness** and relieving pain. It is an alternative **medicine** carried out by trained people called osteopaths. Bones, joints, muscles, ligaments and tendons are moved and massaged by hand with this type of treatment. Osteopathy deals with all kinds of disorders of the skeleton and nervous systems.
The footballer's leg was treated by osteopathy.

otitis media *noun*
Otitis media is an infection of the **middle ear**. This part of the ear becomes inflamed and full of pus, causing very painful earache, a fever and deafness. When someone has a cold, the infection can get into the middle ear through the **Eustachian tube** at the back of the nose. If the pus builds up it may burst the eardrum, which relieves the pain. Antibiotic drugs and draining out the pus are treatments for otitis media.
Otitis media is often caused by virus infections.

outer ear *noun*
An outer ear is shell of skin and cartilage It has a funnel-shaped canal which leads to the **middle ear**. The outer ear is separated from the middle ear by the **eardrum**. Sound waves passes through the canal to the eardrum. Fine hairs and wax inside the outer ear protect the ear from dust and other particles.
The boy punched him on his outer ear.

ovary *noun*
An ovary is a female sex organ inside the **pelvis**. The two ovaries are about the size of a walnut and are attached either side of the **uterus**. The arms of the **Fallopian tubes** of the uterus curve over and meet the ovaries. Ovaries make and store **ova** and produce **hormones** which help to control the **menstrual cycle**.
Hormones cause the ovaries to release eggs.
ovarian adjective

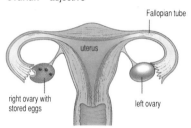

Fallopian tube

uterus

right ovary with stored eggs

left ovary

ovulation *noun*
Ovulation is the release of an **ovum** from an **ovary**. One mature ovum is released half-way through the **menstrual cycle** every month. The egg travels along the **Fallopian tube** from the ovary to the **uterus**. If it is not fertilized, the ovum leaves the body during **menstruation**.
Ovulation is the time when a woman is at her most fertile.

ovum (plural ova) *noun*
An ovum is a female egg that consists of one cell. An ovum has 23 **chromosomes**, half the usual number. A male **sperm** is the only other kind of cell with 23 chromosomes. An ovum is released every month from an **ovary** and travels through a **Fallopian tube** to the **uterus**. When a sperm meets and **fertilizes** an ovum, the chromosomes pair up to make an **embryo**.
An unfertilized ovum passes out of the body during menstruation.

oxygen *noun*

Oxygen is a gas in the air. It has no smell or colour and makes up one fifth of normal air. Humans need oxygen to live. It is absorbed into the **blood** from the air breathed into the **lungs**.
She took a deep breath to get more oxygen to her lungs.

pacemaker *noun*

A pacemaker is part of a **heart**. It is a small area of **tissue** which controls the **heartbeat**. It does this by sending out an electric signal. This tells the heart how fast and how often to pump out **blood**. If the heart's pacemaker fails, an artificial electronic one can be fitted under the skin in the chest. Wires from the pacemaker are connected to the heart.
A pacemaker gave him a longer life.

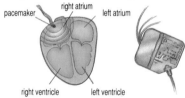

pacemaker   right atrium   left atrium

right ventricle   left ventricle

pain *noun*

Pain is a feeling of being hurt somewhere in the body. **Sensory nerves** carry pain signals to the **brain**. Pain disappears as an injury heals or the illness is treated. Severe pain can be eased with **drugs** called **analgesics**.
Most pain is a sign of injury or illness.

palate *noun*

The palate is the roof of the **mouth**. The hard palate in front is made of bone. The soft palate curves down towards the back of the throat. This part of the palate is made of muscle and ends in a small lump of tissue called the **uvula**.
The palate helps you to swallow food without choking.

palm *noun*
A palm is the inside part of a **hand**. Palms start from the wrist and end at the base of the fingers.
Lines criss-cross the palm of your hand.

pancreas *noun*
The pancreas is a **gland** in the **abdomen**. It is between 12 and 15 centimetres long and is joined to the **small intestine**, behind the **stomach**. Juices from the pancreas help to break down, or **digest**. The pancreas also makes **insulin**, a **hormone** that controls the level of sugar in the **blood**.
He suffered from diabetes because his pancreas made too little insulin.

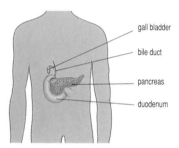

gall bladder

bile duct

pancreas

duodenum

pandemic *adjective*
Pandemic describes any **disease** that affects many people in a large area of the world. A disease that spreads across a smaller area is described as **epidemic**.
Malaria is a pandemic disease, found in most tropical countries.

paralysis *noun*
Paralysis is not being able to move. This may be in part or all of the body. Most paralysis is caused by damage to the **nerves** that tell the **muscles** to move. This can happen through a disease or an injury.
Paralysis may affect a person for a short time, a long time or the rest of their life.

parasite ► **worm**

parathyroid gland *noun*
A parathyroid gland is one of four small organs in the neck. There are two each side of the **thyroid gland** in the front of the neck. Parathyroid glands release a **hormone** that controls the level of calcium in the **blood**.
Parathyroid glands are part of the endocrine system.

parent *noun*
A parent is the mother or father of a child.
Parents provide love, care and training for their children.

patella ► **kneecap**

pathogen *noun*
A pathogen is a substance or any kind of **germ** that causes a disease. Some types of **bacteria** are pathogens because they enter the body and cause infections.
Pathogens in the drinking water caused him to become ill.

patient *noun*
A patient is a word for a person who is ill. People in **hospital** are called patients.
The patients lay on their beds in the hospital ward.

pectoral *adjective*
Pectoral describes anything to do with the **chest**.
He did a lot of heavy weightlifting to develop his pectoral muscles.

pelvis *noun*
A pelvis is a large circle of **bone**. This is the part of the **skeleton** where the legs join the rest of the body. A pelvis is a hollow cradle of bone which protects the organs in the lower **abdomen**. The hipbones and lower backbone fuse together to form the pelvis. The main bones of the pelvis are called the **hipbone**, **sacrum** and **coccyx**. The pelvis of a woman is usually wider than a man's.
The belly dancer rotated her pelvis.
pelvic *adjective*

penicillin   *noun*
Penicillin is a **drug**. It is one of a group of
drugs called **antibiotics**. These kind of drugs
fight harmful **bacteria** in the human body.
Penicillin was first made from the tiny living
things that form mould. The bacteria in the
mould kill certain kinds of bacteria that cause
infections. There are now many penicillins
made from other substances. A few people
are allergic to penicillin.
*A British scientist called Alexander Fleming
discovered penicillin in 1928.*

penis   noun
A penis is a **male** sex **organ**. The penis lies
over the **scrotum** between the legs. The
head of the penis is moistened and covered
by the **foreskin**. **Urine** and **semen** pass out
of the penis through a tube called the
urethra. The penis is full of spongy tissue.
When this fills up with blood, the penis grows
larger, becomes firm and stands on end. The
erection of the penis is an important part of
sexual intercourse.
*During adolescence, a boy's penis grows
larger.*

people in medicine ► page 112

pepsin   *noun*
Pepsin is an **enzyme** which helps to **digest**
food. It is found in the juices made by the
walls of the **stomach**. Pepsin breaks down
the **protein** in food into smaller particles.
*Pepsin is just one of the enzymes that help
us to digest the food we eat.*

peritoneum   *noun*
The peritoneum is a thin **membrane** inside
the body. It lines the inside of the **abdomen**
and wraps and protects all the organs in the
abdomen, such as the **stomach**, the
kidneys and the **liver**. The thickest part of
the peritoneum supports the **intestines** at
the back of the abdomen. It also links the
intestines with the **lymphatic system**.
*A puncture of the peritoneum can be
extremely painful.*

peritonitis   *noun*
Peritonitis is an infection of the **peritoneum**.
This infection causes part of the peritoneum
to become inflamed. A person will feel
severe pain in the **abdomen** and may have a
fever. Peritonitis is usually caused by an
infection of one of the **organs** inside the
abdomen, such as the **appendix**.
*The child developed peritonitis and had to be
rushed to hospital.*

phalanges   *plural noun*
The phalanges are the **bones** of the **fingers**
and **toes**. There are two bones in each of the
thumbs and big toes. The rest of the fingers
and toes have three bones each. A single
bone of the toe or finger is called a phalanx.
*He broke one of the phalanges in his finger
when he shut it in the train door.*

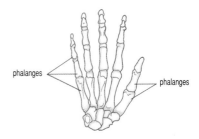

phalanges

phalanges

pharmacy   *noun*
1. A pharmacy is a place where **medicines**
are made and supplied. Most **hospitals** have
their own pharmacy. Pharmacists are trained
to prepare and mix medicines in a pharmacy.
*From his pharmacy, the pharmacist filled
many prescriptions.*
2. Pharmacy is the process of making
medicines to treat illness. Scientists use all
kinds of substances to make **drugs** and
other medicines.
*The pharmaceutical company made pain-
relieving tablets.*
pharmaceutical   adjective

people in medicine *noun*

People in medicine have always searched for ways to treat **illnesses** and **injuries**. Thousands of years ago in India, surgical operations were performed using instruments much like those in use today. The Ancient Chinese developed acupuncture, and used plants as drugs. Over the years, scientists have made discoveries that help them to understand how to prevent illness and give the best treatment.

Modern technology helps people in medicine to treat many kinds of disease and injury.

Edward Jenner 1749–1823
Jenner was a British doctor who discovered vaccination. He found that if someone was injected with mild cowpox germs, they did not catch smallpox.

Andreas Vesalius 1514–1564
Vesalius was a Flemish doctor who made the first accurate anatomical studies of the human body. He did much of his research at his medical school in Italy.

Louis Pasteur 1822–1895
The French scientist Pasteur discovered that bacteria can cause disease. He found that heating killed the bacteria. He also discovered ways of weakening harmful germs so that they could be used as vaccines.

Wilhelm Roentgen 1845–1923
Roentgen was a German physicist who discovered X-rays in 1895. His work encouraged other physicists to study other types of invisible radiation.

Shibasaburo Kitasato 1852–1931
The Japanese scientist Kitasato studied the bacteria that cause tetanus and diphtheria, and also discovered the bacterium that causes bubonic plague. Kitasato founded a laboratory near Tokyo, Japan, where infectious diseases were studied.

Pierre Curie 1859–1906
Marie Curie 1867–1934
The Curies were French scientists who discovered the radioactive substance called radium. Radium was used to treat cancer. After her husband's death, Marie Curie devoted much of her time to the medical use of radioactive substances.

Alexander Fleming 1881–1955
Fleming was a Scottish scientist who discovered penicillin. Penicillin is a mould that can kill germs. This discovery led to the development of antibiotics, which greatly reduce the risk of infection after an injury or surgery.

pharynx *noun*
A pharynx is part of a **throat**. The tube of the pharynx joins the back of the **nose** and **mouth** to the **larynx**. It is a moist and muscular pathway for air and food. The sounds made by the larynx use the space of the pharynx as an echo chamber. The pharynx also contains the **tonsils** and the **adenoids.**
He had a sore throat due to an infection of his pharynx.

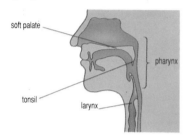

soft palate
pharynx
tonsil
larynx

phlegm *noun*
Phlegm is the thick **mucus** that sometimes collects in the nose and throat. People find they have phlegm when they have a cold.
Her throat was so blocked with phlegm that she could hardly breathe.

phobia *noun*
A phobia is a strong fear of something. This may be of an animal, such as a spider, or it may be a fear of heights or the dark. Agoraphobia is a fear of being outside. Claustrophobia is a fear of being shut in.
Since she was a child, she had had a phobia of mice.

physiology *noun*
Physiology is a science. In physiology, living **cells**, **tissues** and **organs** are studied to see how they work. A person who studies this subject is called a physiologist.
The student studied physiology to learn about the human body.

pigment *noun*
A pigment is a substance that gives colour. The pigment in **blood** is red and a pigment in **bile** gives it a yellowish-green colour. The pigment in skin is called **melanin.**
Freckles are caused by particles of pigment in the skin.

piles ► hemorrhoids

pimple ► spot

pituitary gland *noun*
A pituitary gland is a small **organ** at the base of the **brain**. It is joined to the **hypothalamus**, which is the part of the brain that controls eating, drinking and body heat. The pituitary gland is the main part of the **endocrine system**. It tells all the other glands to work by releasing **hormones**. In a female the pituitary gland also releases hormones which start **childbirth** and **breast-feeding**.
The pituitary gland releases growth hormone that helps to determine how tall we grow.

pivot joint ► joint

placenta *noun*
A placenta is an **organ** in a pregnant female. The unborn baby in the **uterus** is fed through the placenta. The **umbilical cord** links the baby to the placenta, which also takes away waste substances. The placenta is passed out of the **vagina** after **childbirth**.
Food and oxygen pass from the mother to the baby through the placenta.
placental *adjective*

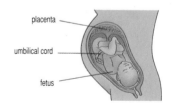

placenta
umbilical cord
fetus

plaque *noun*
Plaque is a substance that builds up to make a covering. **Bacteria** form a hard, scaly layer on teeth known as plaque. This type of plaque can cause gum disease, such as **gingivitis**, or **caries** and tooth decay. A scabby patch of **skin** is also known as plaque. The third kind of plaque in the body is the layer of **fat** that sometimes lines the blood vessels.
His teeth were decaying because he never brushed the plaque off them.

plasma *noun*
Plasma is a part of **blood**. It is the pale yellow fluid that surrounds the tiny **blood cells** and **platelets** that make up blood. Plasma is a solution of water, salts and chemicals. This fluid carries substances, such as **sugar** and **hormones**, to all parts of the body. Plasma also removes waste products. It is possible to give plasma by **transfusion**.
Plasma is the fluid that makes up most of your blood.

plaster cast *noun*
A plaster cast is a mixture of powder and water that sets hard. It is used to support broken **bones** in the legs and arms. The plaster paste is moulded around the **limb** with bandages. It quickly sets into a hard shell which keeps the bones in place until they heal.
The doctor covered her broken leg with a plaster cast.

platelet *noun*
A platelet is a part of blood. Platelets are tiny discs that develop from cells produced in the **bone marrow**. They float in the fluid, or **plasma**, of blood. The billions of platelets in the body help to make the blood **clot**.
When you cut yourself, platelets help to stop the bleeding.

pleura (plural **pleurae**) *noun*
A pleura is one of two layers of **skin** in the **chest**. One layer wraps round the **lung**. The other layer lines the inside of the chest. There is a narrow space between the two layers called the pleural cavity. Fluid in the pleural cavity allows the layers of the pleura to glide past each other every time the lungs move to breathe.
The pleurae are two layers of skin inside the chest cavity.

pleurisy *noun*
Pleurisy is a disorder of one or both **lungs**. The layers of skin that surround the lungs, or **pleurae**, become inflamed. Because they are dry and rough, they rub together instead of gliding smoothly past each other. This causes a sharp pain in the chest, which is worse when a person coughs or breathes deeply. Pleurisy is usually a symptom of another illness, such as **pneumonia**.
Pleurisy made it difficult for her to breathe without pain.

pneumonia *noun*
Pneumonia is a disorder of one or both **lungs**. It happens when harmful germs are breathed into one or both of the lungs and start an **infection**. Pneumonia may develop when a person is already ill, very young or very old. At these times their defence against illness, or **immune system**, is not very strong. Pneumonia may start with the symptoms of a cold or **influenza**. A person may be breathless and feel a pain on one side of the chest.
Antibiotic drugs are used to fight lung infections such as pneumonia.

poliomyelitis *noun*
Poliomyelitis is a **disease**. It is carried by a **virus** that attacks **nerves** in the **brain** and **spinal cord**. The virus is catching and can be carried by people with no **symptoms**. The first signs of poliomyelitis are a **fever**, sore **throat** and **headaches**. A small number of people may then develop a serious form of poliomyelitis with **muscle** pain and a stiff neck. This can lead to **paralysis,** or loss of movement in part or all of the body. Most people recover from the paralysis, but in some cases paralysis may remain.
People can be protected from poliomyelitis by a vaccine given in childhood.

polyp *noun*
A polyp is a lump that forms in moist **skin**. Polyps are usually **benign**, or harmless, but they may cause a blockage. Common places for polyps to grow are in the **nose, ear, stomach** and **intestine**. They may also develop in the **uterus**.
Polyps that cause a blockage or become infected may be removed.

pore *noun*
A pore is a tiny opening in the **skin**. **Sweat** passes through pores to the surface of the skin.
Washing with cool water helped to close his pores.

posterior *adjective*
Posterior describes the back of any part of the human body. The opposite of posterior is **anterior**, which means front.
The spine is a posterior part of the body.

posture *noun*
Posture is the position of the **body**. Good posture is important when walking, standing or sitting. The **back** should be straight, **shoulders** relaxed, **head** held up and **stomach** held in.
Bad posture with hunched shoulders and a curved back strains muscles and causes pain.

pregnancy *noun*
Pregnancy is the time before a baby is born. Pregnancy starts at **conception** and carries on until **childbirth**. This lasts about nine months, or 40 weeks. The first sign of pregnancy is a missed period, or time of **menstruation**. After three months, the mother's **abdomen** starts to swell as the baby grows inside the **uterus**.
During the fifth month of her pregnancy, she could feel the baby kicking inside her.

premolar ► **tooth**

preventive medicine *noun*
Preventive medicine is a kind of **health** care. Giving **vaccinations** and having regular **check-ups** prevent illness. Preventive medicine also teaches people what harms the body and how to protect it. Learning all about road safety, first aid and how to live a healthy lifestyle are all kinds of preventive medicine.
Preventive medicine can help people to avoid becoming ill.

probe *noun*
A probe is a thin metal rod with a blunt end. **Doctors** use probes to explore openings in the body, such as a nostril. Probes are also used to look into wounds to find a foreign body, such as a bullet. **Dentists** use probes when they are checking a person's teeth.
The doctor used a probe to unblock the girl's tear duct.
probe *verb*

prognosis *noun*
A prognosis is made by a **doctor** after seeing a **patient**. First the doctor finds out the name of the **illness**, or makes a **diagnosis**. The doctor is then able to tell how the illness may develop and what the patient can expect to happen. This is the prognosis.
The prognosis was that the fever would end in two or three days.

prostaglandin *noun*
A prostaglandin is one of a group of **hormones**. These hormones are found in many tissues and fluids in the body. They activate the body in several different ways. Some kinds of prostaglandin control the acid in the **stomach** juices. Others make the **uterus** contract during **childbirth**. The chemicals in prostaglandins can be used as **medicines.**
Prostaglandins cause a pregnant woman to start labour.

prostate gland *noun*
A prostate gland is a small **organ** in a male. It is the size of a walnut and lies around the **urethra**, the tube that carries **urine** from the **bladder**. The prostate gland releases a fluid that is part of **semen**. In older men, this gland may swell and squeeze the urethra. This stops the flow of urine. When this happens, the prostate gland can be removed.
The enlarged prostate gland was the reason for his bladder infection.

protein *noun*
Protein is a substance made of **amino acids**. The body needs protein to build and repair **cells**. Different kinds of protein keep **bones**, **blood** and many other parts of the body healthy. Some proteins are made by the body. Others need to be eaten in **food**. Meat, fish, eggs, milk and cheese all contain important proteins the body needs.
Children and pregnant women need to have a diet rich in protein.

psoriasis *noun*
Psoriasis is a disorder of the **skin**. Small, silvery scales appear on the skin, with red patches underneath. Sometimes, large areas of psoriasis may cluster on the head, elbows and knees. These areas can be sore, especially when a person is under **stress**, or becomes upset. Psoriasis is not catching and usually runs in families, or is **hereditary**. Sunlight is a good treatment for psoriasis.
The ointment helped to clear up the psoriasis on her elbows.

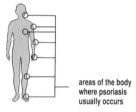

areas of the body where psoriasis usually occurs

psychiatry *noun*
Psychiatry is a way of treating all kinds of mental illness. Doctors who give this kind of medical care are called **psychiatrists**. Some kinds of drug are helpful in treating disorders of the **mind**. A psychiatric **hospital** looks after people who need this kind of care.
The psychiatry students learned how drugs, discussion and therapy might help the patients.

puberty *noun*
Puberty is a stage of growth in a person. It happens when a child begins to develop into an **adult**. Puberty lasts between two and six years. The glands of the body become very active and produce hormones. During puberty, girls develop **breasts** and start to **menstruate**. The **penis** and **testes** of a boy grow larger and start to produce **semen**, and his voice becomes deeper. Both sexes grow hair under the arms and around the sexual organs between the legs.
Puberty can start any time between 11 and 17 years of age.

pubic hair   *noun*

Pubic hair is a kind of hair that starts to grow on boys and girls during **puberty**. It is the thick, short, curly hair that grows around the sexual organs between the legs. A few hairs appear at first around the penis or vulva. More hairs grow and form an upside-down triangle shape.

Pubic hair grows between the legs to protect the outer sexual organs.

pubis   *noun*

A pubis is one of two **bones** that form the front ring of the **pelvis**. The pelvis is the circle of bone where the legs meet the rest of the body. The pubic bones curve round the front of the pelvis and meet in the middle.

The two bones of the pubis meet at the pubic arch.

pulmonary   *adjective*

Pulmonary describes anything to do with the **lungs**. **Asthma** is a pulmonary disease.

The mild pulmonary infection could easily have developed into pneumonia.

pulse   *noun*

A pulse is a throb of pressure in the flow of **blood**. The **heart** pumps blood into the arteries, to be carried around the body. The blood flows in a wave every time the heart beats. These surges can be felt as a pulse in the wrist and neck. A pulse that is weak, too fast, or too slow may be a sign of some kind of **illness**.

He was nervous about the interview, so his pulse beat faster.

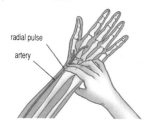

radial pulse

artery

pupil   *noun*

A pupil is part of an **eye**. It is the round black opening in the middle of the **iris**, the coloured ring of an eye. Light enters the eye through the pupil. The amount of light let in is controlled by **muscles** in the iris. This makes the pupil look bigger in dim light and smaller in bright light.

His pupils appeared to grow larger as he moved into the shade.

pus   *noun*

Pus is a thick, yellow or green liquid made by the body. It collects around **infection** in **tissue**. This may cause a painful, red lump, such as an **abscess**. Pus is a sign that the body's defence, or **immune system,** is at work. Pus is formed from dead **white blood cells** with the **germs** they were fighting, dead tissue and fluid. Pus needs to be drained and cleaned from an infected area.

The wound had become infected, and the bandage was covered with pus.

quarantine   *noun*

Quarantine is a period of time. It is the time after a person is in contact with a **contagious disease**. This kind of disease takes days, or weeks to develop. A person must stay away from other people during this time in case they pass on the illness. The quarantine period for mumps is 28 days.
A yellow flag was used on a boat to show that it was under quarantine.

rabies   *noun*

Rabies is a **disease**. It is caused by the bite of an animal infected with the rabies **virus**. Wild animals such as bats, foxes and cats spread the illness in some countries. The first signs are a **fever** followed by excited behaviour. A person froths at the mouth, the throat closes up and they cannot bear to drink water. Someone who has been bitten by an animal with rabies needs medical treatment at once. Once **symptoms** develop, a person dies within a week.
A person who has been exposed to rabies needs to have a series of six injections of vaccine over a one-month period.

radiography   *noun*

Radiography is a way of looking inside the body. This is usually by making **X-rays**. These are pictures of bones and organs inside the body. Looking at pictures of the inside of the body helps doctors to make a **diagnosis**. They can see how a bone is broken or if there is something wrong with an **organ**.
Radiography was used to find out exactly where the bone was broken.

radiotherapy   *noun*

Radiotherapy is a way of treating many kinds of **cancer**. This treatment uses invisible waves of energy called radiation to destroy cancer **cells**. Radiotherapy is often used after an operation to remove a **malignant** growth. The treatment kills any cancer cells left behind.
Sometimes, a patient's hair falls out after a course of radiotherapy.

radius   *noun*
A radius is a **bone** in the **arm**. It is the front
bone of the forearm. The top end forms part
of the **elbow** and the bottom end is part of
the **wrist** joint on the **thumb** side of the
hand. The radius can move over and across
the **ulna**, the other bone in the forearm. This
allows the **palm** of the hand to face up or
down by moving the wrist.
Attached to the radius and the ulna are
19 muscles which move the wrist and the
fingers.

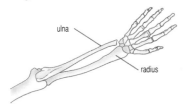

ulna

radius

rash   *noun*
A rash is a skin reaction or infection. Rashes
are usually red pimples that itch. The spots
may be raised, turn into blisters or break
open. Rashes can be caused by a skin
irritation or an **allergy**. A rash may also be a
symptom of another illness, such as
measles or chickenpox.
When the boy fell into the nettles, he had a
rash all over his legs.

rectum   *noun*
A rectum is a tube in the body. It is the end
section of the large **intestine** where **feces**
are stored. The rectum is about 12
centimetres long and reaches from the **colon**
to the **anus**. It passes through the **pelvis** and
lies behind the **bladder.** When feces fill up
the rectum, a person feels the need to go to
the toilet.
The colon is joined to the anus by the
rectum.

red blood cell ► **blood**

reflex ► **motor reflex**

reflex ► page 121

renal   *adjective*
Renal describes anything to do with a
kidney.
Blood flows into the kidney through the renal
artery.

reproduction   *noun*
Reproduction is the process of making
babies. For reproduction, a **female** and a
male must be **fertile** and have **sexual**
intercourse. Fertilization of the ovum is
followed by **pregnancy** and **childbirth**.
Without reproduction, the human race would
die out.

respiration ► page 122

retina   *noun*
A retina is part of an **eye**. It is a layer of
special **cells** that lines the back of an
eyeball. Light is focused onto the retina
through the **lens**. Cells called **rods** pick up
light and other cells called **cones** pick up
colour. The messages of light and colour are
sent along the **optic nerve** to the **brain**.
Damage to the retina can cause blindness.

rheumatic fever   *noun*
Rheumatic fever is a **disease**. It is most
often found in children between 5 and 15
years of age. Rheumatic fever is caused by
an unusual reaction to a throat infection. Two
weeks after a sore throat, the joints become
swollen, red and painful. There may be a
fever, rash, and chest and stomach pains.
Most people recover completely from
rheumatic fever, although a serious attack
may damage the heart.

rheumatism   *noun*
Rheumatism is pain, stiffness and
inflammation in muscles and joints. Many
elderly people suffer from rheumatism.
His rheumatism was relieved by aspirin.

reflex *noun*

A reflex is an automatic response made by the body. This can be a **movement**, as in a **motor** reflex. Other kinds of reflex action can develop through memory. The brain remembers how to respond to certain signals or situations. This is called a conditioned reflex. Hearing an ice-cream van might make someone's mouth water. This is a conditioned reflex.

Blinking when something suddenly comes too close to the eyes is a motor reflex.

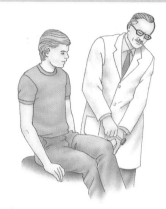

Doctors test the knee reflex to make sure the nervous system is intact. To test the knee reflex, the patient sits with one knee crossed over the other. The doctor taps the knee just below the kneecap.

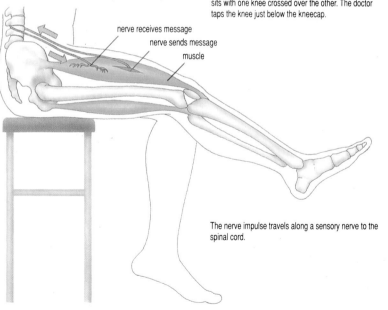

nerve receives message

nerve sends message

muscle

The nerve impulse travels along a sensory nerve to the spinal cord.

respiration   *noun*

Respiration is the process by which the body obtains **oxygen** and gets rid of a waste gas called **carbon dioxide**. Humans obtain oxygen by breathing air. Air is drawn in to the **lungs**, where oxygen is taken into the **blood** and circulated to every **cell** in the body. Carbon dioxide leaves the body when a person breathes out. Respiration also means breathing.

During respiration, an adult breathes 15 or 20 times a minute.

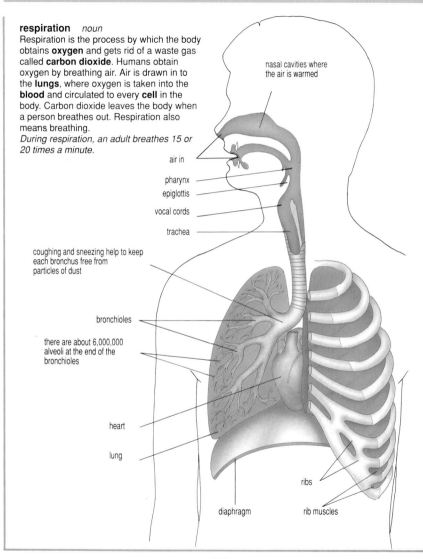

nasal cavities where the air is warmed

air in

pharynx

epiglottis

vocal cords

trachea

coughing and sneezing help to keep each bronchus free from particles of dust

bronchioles

there are about 6,000,000 alveoli at the end of the bronchioles

heart

lung

ribs

rib muscles

diaphragm

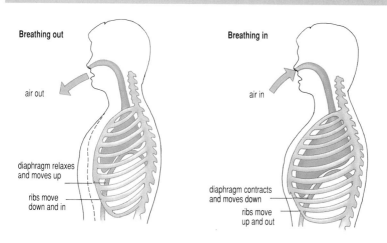

Breathing out

air out

diaphragm relaxes
and moves up

ribs move
down and in

Breathing in

air in

diaphragm contracts
and moves down

ribs move
up and out

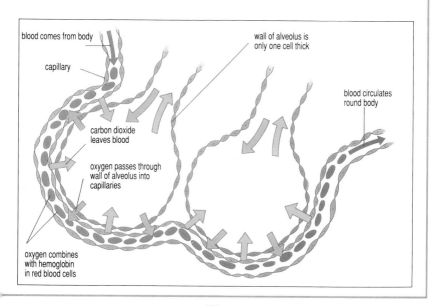

blood comes from body

capillary

wall of alveolus is
only one cell thick

blood circulates
round body

carbon dioxide
leaves blood

oxygen passes through
wall of alveolus into
capillaries

oxygen combines
with hemoglobin
in red blood cells

rib   noun
A rib is a thin, curved **bone** in the **chest**.
There are 12 pairs of ribs in the body. Each
pair of ribs is joined to the **spine** at the back.
The first seven pairs join the **breastbone** at
the front. Three more pairs are connected to
the breastbone with **cartilage**. The last two
pairs only curve half way round and are
called floating ribs.
*He could feel his ribs when he ran his hand
down his side.*

rickets   *noun*
Rickets is a **disease**. It happens when
children do not have enough **vitamin** D. The
skin makes vitamin D when a person is in
sunlight. It is also found in some foods, such
as fish and eggs. Rickets affects the **bones**
in a child's body. It stops them growing
properly and makes them soft.
*Cheese is rich in vitamin D, and can help to
prevent rickets.*

rod ► eye

root canal   *noun*
A root canal is part of a **tooth**. It is the
passage through the root of a tooth buried in
the **gum**. Root canals connect the soft pulp
inside teeth with nerves and a supply of
blood. If the inside of a tooth dies, the root
canal must be cleaned and filled by a
dentist.
*He needed root canal surgery to treat his
toothache.*

rubella   *noun*
Rubella is a **contagious disease**, caused by
a **virus**. It is often called German measles.
The symptoms of rubella are a **headache**,
fever and sore throat followed by a **rash** of
small red spots. Rubella is contagious from a
week before until five days after the rash
appears. If rubella is caught by a woman in
the first three months of pregnancy, it can
harm the baby she is carrying.
*Children are usually vaccinated against
rubella so they will never catch the disease.*

sac   *noun*
A sac is a pouch of **tissue**. Sacs line hollows
or surround spaces in the body. They may
also contain fluid, organs or other tissue.
There are many kinds of sacs in the body,
such as the tiny air sacs in the lungs called
alveoli.
The scrotal sac contains a man's testes.

sacro-   *prefix*
Sacro- is a prefix that refers to anything to do
with the **sacrum**.
*The sacroiliac joint connects the sacrum to
the ilium.*

sacrum   *noun*
A sacrum is part of a **spine**. It is five bones,
or **vertebrae**, of the lower back fused into
one triangular shape. The sacrum forms the
back part of the **pelvis**, with the hipbones
used either side. The bottom of the sacrum
joins the tailbone, or the **coccyx**. This part of
the pelvis takes the weight of the body from
the rest of the spine.
Cartilage connects the sacrum to the coccyx.

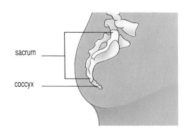

sacrum

coccyx

saline   *noun*
Saline is a fluid. It is made by mixing a tiny amount of a salt called sodium chloride in water. This makes a fluid very like the fluid called **plasma**, in blood. Saline can be injected into the body to replace plasma.
After the operation, the patient was connected to a saline drip.

saliva   *noun*
Saliva is a watery fluid. It is made by the **salivary glands** and moist skin in the mouth. Saliva mixes with **food** to make it easier to swallow. The **enzyme** and salts in saliva start to break down food as it enters the mouth. Saliva enables people to taste flavours in food and helps to keep the mouth clean.
The sight of the chocolate cake made his mouth fill with saliva.

salivary gland   *noun*
A salivary gland releases a fluid called **saliva**. There are three main pairs of salivary glands in the mouth. These are inside the cheeks, at the back of the mouth, below the ears and under the tongue.
The salivary glands beneath the tongue are known as the sublingual glands.

salmonella   *noun*
Salmonella is a **bacterium**. If a person eats food infected with salmonella, they become ill within two days. The bacteria travel to the **intestine** and cause pain, vomiting and diarrhea. People with salmonella poisoning need to drink plenty of fluids.
Salmonella poisoning can be serious in young children and elderly people.

sarcoma   *noun*
A sarcoma is a type of **cancer**. Sarcomas form in **connective tissue** such as **cartilage**, **muscle** or **bone**. The growths are **malignant**, and can spread to other parts of the body. Sarcomas are usually treated by being surgically removed.
She had radiation and chemotherapy after the sarcoma was removed.

scab   *noun*
A scab is a hard crust of dried **blood** and **pus**. Scabs form over a wound. This protects the skin underneath while it is healing. Scabs fall off when the **skin** has healed.
He picked the scab on his knee until it bled.

scabies   *noun*
Scabies is a **skin disease**. It is caused by a tiny mite that burrows under the skin. This makes the skin break out in a red, pimply **rash** with a burning itch. Common places for scabies are the hands and around the sexual organs. The scabies mite is easily passed by touching an infected person.
Scabies is treated by painting on a lotion that kills the mites.

scabies mite

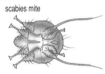

scald   *noun*
A scald is a **burn** caused by hot liquid or vapour, such as boiling water or steam.
She plunged her hand into cold water to stop the scald becoming worse.

scalp   *noun*
A scalp is the **skin** and **hair** that covers and protects the **skull**. The layer of **muscle** and other **tissue** underneath this area of skin is also part of the scalp.
Headlice made the girl's scalp itch.

scan   *noun*
A scan is a way of looking inside the body. A machine called a **scanner** uses radiation, sound or magnetic forces to make a picture of the **organs** inside the body. All these forces are able to pass through the skin without hurting the body. Computers linked to the machine build up pictures of an organ.
He had to have a scan to see if he had gallstones.

scanner *noun*
A scanner is a machine that makes a **scan**, which is a picture of the body inside the skin. Scanners use sound, radiation or magnetic forces that can pass through the body without hurting any part of it. They are linked to computers which build up pictures of every part of an **organ**. An **ultrasound** scanner is able to build up a picture of a baby inside a pregnant woman on a computer screen.
The scanner could show whether her unborn child was growing properly.

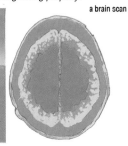
a brain scan

scapula (plural **scapulae**) *noun*
A scapula is a part of a **shoulder**. The scapula, or shoulder blade, is a flat triangle of **bone** that forms the back of the shoulder joint. The **collarbone** links the scapula to the **sternum** at the front of the body. Muscles attached to the scapula help the arm and shoulder to move.
As the girl moved her arm, her scapula moved with it.

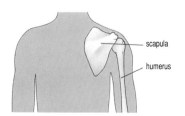
scapula
humerus

scar *noun*
A scar is the mark left after a **wound** heals. Scars from most small injuries heal and fade to a thin, white line. Scars that form into thick, red, lumpy **tissue** can usually be removed by surgery.
He had a long scar on his arm where he fell on the glass.

scarlet fever *noun*
Scarlet fever is a childhood **disease**. It is **infectious** and spreads easily from one child to another. The first signs are a sudden fever and sore throat, headache, sickness and diarrhea. A fine pink **rash** spreads all over the body from the neck and chest. The tongue is coated white and also covered in red spots. The symptoms may only last for a few days, although a person may still be infectious for three weeks.
Scarlet fever is treated with antibiotics.

schistosomiasis ► bilharzia

scrotum *noun*
A scrotum is part of a male body. The scrotum is a loose bag of skin that contains the **testes**. It hangs below the **penis** between the legs. During **puberty**, the testes enlarge and drop lower and the scrotum gets bigger. One testis hangs lower than the other. **Muscles** in the scrotum pull the testes up towards the body if they get cold. This is to keep the **sperm** inside at the right temperature.
The scrotum keeps the testes at the right temperature.

scurvy *noun*
Scurvy is a **disease**. It happens when a person does not have enough **vitamin** C in their diet. Vitamin C is found in fresh fruits and vegetables. The first sign of scurvy is swollen, bleeding **gums**, followed by bleeding under the skin. This may lead to a blood disorder called **anemia**.
People with scurvy soon recover by adding vitamin C to a balanced diet of food.

seasickness   *noun*
Seasickness is a kind of motion sickness
caused by the rocking of a boat on the sea.
The movements upset the organs of
balance, or **semi-circular canals**, in the
ear. Someone with seasickness feels sick
and **vomits**. There are medicines which help
to stop people feeling ill when they travel.
Children often grow out of seasickness.

sebaceous gland   *noun*
A sebaceous gland is a part of **skin**. These
tiny glands open into **hair follicles** and
release an oily substance called **sebum**.
*A blocked sebaceous gland can become
infected and swell up like a boil.*

sebum   *noun*
Sebum is a thick, oily substance. It is made
by the **sebaceous glands** that surround and
feed each hair root. Sebum is released into
the **hair follicles** under the **skin**.
*Sebum keeps hair soft and clean if dust and
dirt is brushed out.*

secrete   *verb*
To secrete is to release a substance. **Glands**
in the body, such as salivary glands, secrete
fluids through tubes. Some substances are
also secreted straight into the **bloodstream**,
such as **hormones** from the **ovaries**. Any
substance or fluid made and secreted by a
gland is called a secretion.
*A very nervous person secretes sweat from
the palms of the hands.*

semen   *noun*
Semen is a thick, creamy fluid. It is made in
the **testes** of a **male** during **puberty** and
released through the **penis**. Each release of
semen contains millions of **sperm** that swim
in other fluids made in the testes. The sperm
are so tiny that one release, or **ejaculation**,
of semen is only about two teaspoonsful of
liquid. Semen is an essential part of
reproduction and **sexual intercourse.**
*A man's semen is necessary to fertilize a
woman's ovum.*

semicircular canal   *noun*
A semicircular canal is part of the **inner ear**.
The semicircular canals are three tubes filled
with fluid. These are arranged as loops that
join the rest of the inner ear, or the **cochlea**.
Semicircular canals are the main organs of
balance. The fluid in the loops moves as the
body moves and sends a message to the
brain. The brain then instructs the body to
move in the right way to keep in balance.
*Infection in the semicircular canal can make
you lose your balance.*

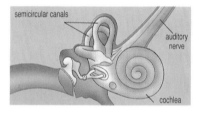

sense   *noun*
The senses are sight, hearing, touch, taste
and smell. The **eyes**, **ears**, **skin**, **mouth** and
nose are the sense organs.
*He used his sense of smell to tell what kind
of flower it was.*

sensory nerve ► **nerve ending**

septic   *adjective*
Septic describes an infection that forms **pus**.
This is caused by germs invading the body,
often through a cut or wound in the **skin**.
*He needed antibiotics to clear up the
infection in the septic cut.*

septicemia   *noun*
Septicemia is a **blood disease**. It happens
when germs get into the **bloodstream** and
poison the blood. This may cause infections,
such as **abscesses**. The first signs of
septicemia are red streaks leading away
from a wound on the skin, fever and chills.
*A person with septicemia needs medical
treatment at once.*

skeleton *noun*

The skeleton is the framework of all the **bones** in the body. There are over 200 bones in an adult skeleton. These are all linked by **joints** and held in place by **ligaments**. **Tendons** attach layers of **muscle** to the bones. Muscles control all the moving parts of a skeleton. The skeleton supports the body and protects all the **organs** inside.

A baby's skeleton contains about 350 bones, many of which fuse together as the baby grows.

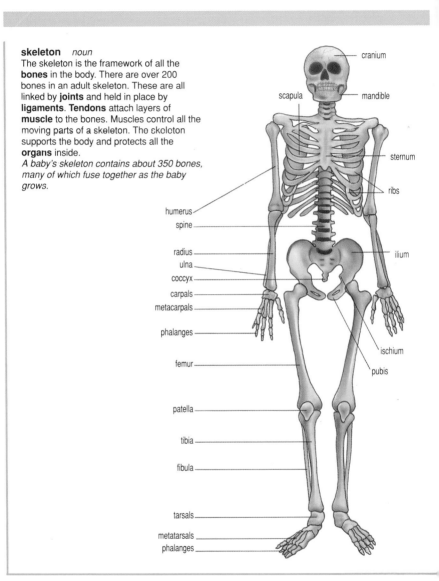

cranium

mandible

scapula

sternum

ribs

humerus

spine

radius

ulna

coccyx

carpals

metacarpals

phalanges

ilium

ischium

pubis

femur

patella

tibia

fibula

tarsals

metatarsals

phalanges

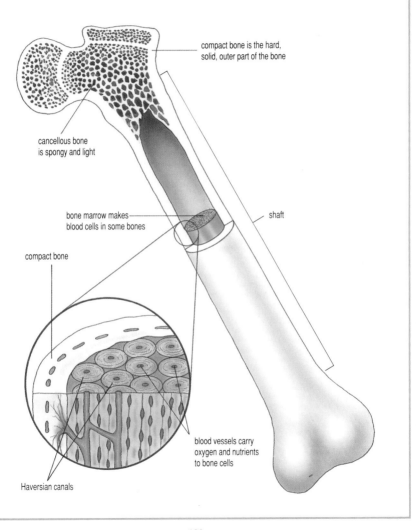

compact bone is the hard, solid, outer part of the bone

cancellous bone is spongy and light

bone marrow makes blood cells in some bones

shaft

compact bone

Haversian canals

blood vessels carry oxygen and nutrients to bone cells

septum (plural **septa**) *noun*
A septum is a wall of thin **tissue**. Septa divide one part of an organ from another, such as the septum in the **nose** that separates the two nostrils. Other septa in the body include those that divide the chambers of the **heart**, the **lobes** of the **brain** and the two halves of the **scrotum**.
She was born with a hole in the septum of her heart.
septal *adjective*

serum (plural **sera**) *noun*
Serum is part of **blood**. It is the clear fluid that is left after a blood **clot** forms. Serum is similar to **plasma**, the other fluid in blood, but it does not contain the substances which make blood clot. Serum can be used as a medicine to help treat some kinds of disease. It can also be used to give people **immunity** from certain diseases.
The explorers took plenty of snake bite serum with them.
serous *adjective*

sex *noun*
Sex refers to whether a person is **male** or **female**. The sex of a baby is decided when a baby is first made, at conception. It depends on which types of **chromosome** join together during conception.
Nowadays you can find out the sex of a baby before it is born.
sexual *adjective*

sexual intercourse *noun*
Sexual intercourse is a part of **reproduction.** The sexual **organs** of a man and a woman meet during sexual intercourse. The **penis** enters the **vagina** and the two people move their bodies together, usually until **semen** is released from the penis. **Sperm** in the semen swim up into the **uterus** and the **Fallopian tubes**. If a sperm meets a female **ovum** on the way from the **ovary** to the uterus, **conception** takes place.
If men and women did not have sexual intercourse, no more babies would be born.

shin *noun*
A shin is part of a **leg**. The shin is the front part of each leg from the **knee** to the **ankle**.
During a football match, the players are often kicked in the shin.

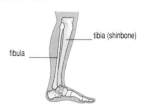

shinbone ► **tibia**

shivering *noun*
Shivering is the way the body shakes and trembles when it is very cold. If the **jaw** shivers, the **teeth** 'chatter' or clack together and make a noise. Shivering is caused by **muscles** which tighten, or contract, very quickly, one after the other. This helps to warm the body if a person is cold, has a shock, or feels afraid.
She was shivering with fear after she thought she saw a ghost.
shiver *verb*

shock *noun*
Shock is an illness that causes the **blood pressure** to drop and not enough blood circulates round the body. Shock may happen when a person loses a lot of blood, has an accident or a severe emotional upset. It can also be a **symptom** of other illness. A person with shock has cold, moist, pale skin. They have a weak and rapid **pulse**, dry mouth and unusual breathing rate. Shock needs to be treated by a doctor as soon as possible.
Everyone involved in the car crash had to be treated for shock.

short-sightedness ► **myopia**

130

shoulder   *noun*
The shoulder joins the **arm** to the rest of the body. Each shoulder has a ball and socket **joint** where the bones meet. The shoulder joint is between the upper arm bone, or **humerus**, and the shoulder blade, or **scapula**. The **collarbone** supports the front of the shoulder. Strong **muscles** and **ligaments** hold the shoulder in place and allow it to move in all directions.
His shoulder ached from throwing the ball so much.

sight   *noun*
Sight is a **sense**. It is being able to see light, colour, shape and size through the **eyes**, the organs of sight. An object is seen at a slightly different angle by each eye. The **brain** makes one complete picture of the object using the information from both eyes. This is called stereoscopic vision. Lack of sight is called **blindness**.
If you have good sight, you do not have to wear glasses.

sinew ► tendon

sinus   *noun*
A sinus is a space, or **cavity,** in the body. The sinuses in the **skull** are lined with a moist skin and lie in the bone behind the nose. These air-filled spaces help to make the skull lighter in weight. They act as echo chambers for the **voice**. Sinuses also warm the air breathed in through the **nose**.
When sinuses become blocked, it is difficult to breathe.

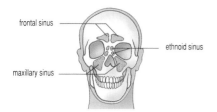

frontal sinus

ethnoid sinus

maxillary sinus

skeleton ► page 128

skin ► page 132

skin blemish   *noun*
Skin blemishes are small, inflamed **swellings** on the **skin**. Skin blemishes happen when the openings, or **pores**, in the skin become blocked. **Sebum**, the oil made by the skin, builds up behind the blockage, and the skin becomes inflamed. Squeezing skin blemishes can let in **germs** and cause **infection**. People often have skin blemishes during **puberty**. This is because the skin makes more oil during this time. Skin blemishes are also called spots or pimples.
Her mother told her not to touch the skin blemishes.

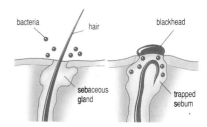

bacteria

hair

blackhead

sebaceous gland

trapped sebum

skull   *noun*
The skull is the bony box that forms the **head**. It is also called the cranium and is made up of 22 **bones**. The part of the skull that covers the **brain** is made from strong, flat bones shaped like a helmet. The bones of the **face** have hollows for the eyes, nose and ears. Some of the facial bones have air-filled **cavities** called **sinuses**. The only bone of the skull that moves is the **jawbone**.
The bones in a new-born baby's skull are separated by cartilage.

sleep ► page 134

skin *noun*

Skin is the outer covering of the body. It is made of tough **tissue** and protects the inside of the body from germs and injury. The outer layer of skin is called the **epidermis**. The middle layer is called the **dermis**. The inner **subcutaneous** layer contains fat. Hair, sweat and blood vessels in the skin help to keep the body at the right **temperature**. Skin also contains nerve cells that send messages of heat, cold and pain to the brain.

The skin is the largest organ in the body.

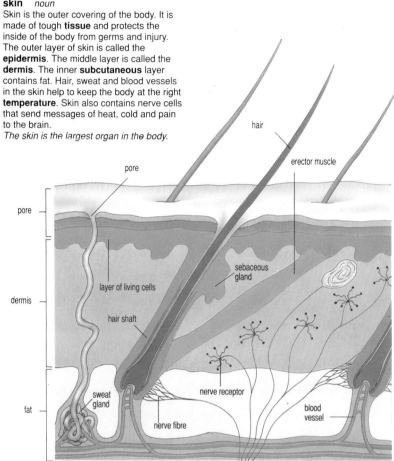

hair

erector muscle

pore

pore

sebaceous gland

layer of living cells

dermis

hair shaft

nerve receptor

sweat gland

fat

blood vessel

nerve fibre

To cool down

To warm up

heat loss

hair

sweat

erector muscle

sweat gland

blood vessels

Blood vessels widen in order to carry more blood near the skin's surface. Heat is given off. Hairs lie flat to ensure no air is trapped between them. Sweat glands produce more sweat. Sweat contains water, salts and other waste matter. Sweat evaporates when it reaches the surface of the skin, cooling the body.

Blood vessels near the surface of the skin become narrower. Blood flows through blood vessels that lie deeper under the skin. Little heat is lost. Erector muscles contract to make hair stand on end. This traps warm air between the hairs to act as insulation. The skin has goose-pimples.

Fingerprints
No two people in the world have exactly identical fingerprints. The patterns on the skin remain the same throughout a person's life. Fingerprints are formed by ridges in the epidermis.

sleep *noun*

Sleep is a time of rest during which a person is not conscious. During sleep, the **heartbeat** and the **breathing** slow down, and the **muscles** relax. Sleep restores **energy** to the body and the **brain**. Most adults sleep for about seven or eight hours a night. Most people have periods of light and deep sleep during the night. People **dream** while they sleep, but often do not remember their dreams.

Growing children need more sleep than adults.

Waves of electrical activity in the brain can be measured by an electroencephalograph machine. The machine prints out the activity as a pattern.

During deep sleep, the eyes remain still. The body is relaxed. Brain waves are large and slow.

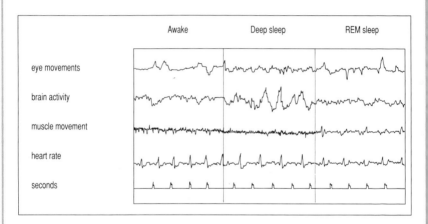

	Awake	Deep sleep	REM sleep
eye movements			
brain activity			
muscle movement			
heart rate			
seconds			

People dream during light, or Rapid Eye Movement (REM), sleep. The eyes flicker under the eyelids. REM brain activity shows smaller, faster waves, in a similar pattern to that when awake.

A sleeping body changes position at least twelve times during a night's sleep.

sleeping sickness   *noun*
Sleeping sickness is a **disease** found in
Africa. It is caused by the bite of a tsetse fly
carrying the sleeping sickness **germ**. Two
weeks after a bite, **fever** and swollen glands
develop. The germs eventually attack the
brain. A person then becomes very sleepy,
falls into a coma and usually dies.
Sleeping sickness can be cured if treated
before the brain is affected.

small intestine   *noun*
The small intestine is part of the **digestive**
system. It is about seven metres long and
coiled inside the **abdomen**. This tube is
divided into three parts called the
duodenum, **jejunum** and **ileum**.
Enzymes secreted by the small intestine help
to digest food.

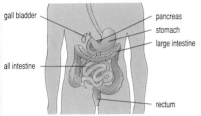

gall bladder — pancreas
— stomach
— large intestine
all intestine —
— rectum

smallpox   *noun*
Smallpox is a **disease**. It is the first disease
to be completely wiped out by a **vaccine**.
The **virus** that caused smallpox has died out,
so there is no-one left to spread the disease.
The symptoms of this illness were a **fever**
and **rash** of red **pimples** over the body.
These filled with **pus** and formed **scabs**.
Scars were left behind after the scabs had
fallen off.
Many people died or were left scarred
through smallpox.

smear   *noun*
A smear is a small amount of tissue or other
substance that is put on a slide for
examination under a microscope.
The smear showed that she had an infection.

smell   *noun*
Smell is a sense. It is being able to pick up
different scents in the air, such as the smell
of food. Special cells in the back of the **nose**
send the message of smells to the **brain**.
Animals have a better sense of smell than
humans, who can only pick up seven main
kinds of scents.
The cerebrum is the part of the brain that
controls smell.
smell   *verb*

sneeze   *noun*
A sneeze is a **reflex action**. A sudden gust
of air is blown out of the **mouth** and **nose**
with great force. This happens when dust or
any other substance irritates the moist skin
inside the nose. A sneeze is the body's way
of cleaning out the passages of the nose.
The loud sneeze startled everyone.

socket   *noun*
A socket is a hollow shape that fits around a
ball shape. The **legs** and **arms** join the rest
of the body with a ball and socket **joint**. The
ball end of one bone fits into the socket end
of another. The **eyeballs** fit into sockets in
the **skull** called **orbits**.
A dislocated bone is one that has come out
of its socket.

solar plexus   *noun*
The solar plexus is a network of **nerves** that
lies at the back of the **abdomen** behind the
stomach. Nerves from the solar plexus
branch out to the rest of the body.
A hard blow to the solar plexus is painful.

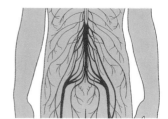

spasm *noun*
A spasm is a movement of a **muscle**, or group of muscles. The muscle suddenly tightens and stays taut. The part of the body seized by a spasm locks into a stiff position.
A spasm may be a simple cramp.

speech *noun*
Speech is the act of speaking. People use speech to talk to each other in many different languages. The sounds of speech are made in the **larynx** and formed by different movements of the **tongue**, **lips** and **jaws**. Children learn speech by hearing and copying other people.
His speech was slurred because he was tired.

sperm *noun*
A sperm is a male **sex** cell. Millions of sperm are made in the **testes** and released through the **penis** in **semen**. Each sperm has a head, a neck and a tail. When sperm are released in a woman's body, they swim up into the **uterus** and **Fallopian tubes**. If one sperm meets a female **ovum**, both cells join together and **fertilization** takes place.
A sperm looks like a tiny tadpole.

sphincter *noun*
A sphincter is a strong ring of **muscle** around an opening in the body. This kind of muscle is able to open and close an opening. The sphincter in the **anus** allows **feces** to pass out of the body.
Sphincter muscles close the opening from the bladder.

spina bifida *noun*
Spina bifida is a disorder of the **spine**. It is a **congenital** disorder, which means that a person is born with it. Babies that are born with spina bifida may have one or more **vertebrae** missing from their spine. Part of the **spinal cord** sometimes sticks out of the spinal column. When the spinal cord is damaged, spina bifida causes **paralysis** in the legs.
Spina bifida can be detected before a baby is born.

spinal cord *noun*
The spinal cord is part of the **nervous system**. It is the dense bundle of **nerves** that runs down the middle of the spine from the **brain**. Nerves in the spinal cord branch out and connect the brain with every part of the body. Messages to and from the body travel through the spinal cord.
If the spinal cord is damaged, paralysis can result.

spinal nerve *noun*
The spinal nerves are the nerves that form the **spinal cord**. There are 31 pairs of spinal nerves which branch out from the **spine** to the rest of the body. Each nerve is joined to the spinal cord by two roots. The front root is a **motor nerve**. This carries messages from the **brain** to the **muscles** and tells them to move. The back root is a **sensory nerve**. This receives messages from the body and sends them to the brain.
Each pair of spinal nerves corresponds with one of the vertebrae.

spine ► page 137

spleen *noun*
A spleen is an **organ** in the **abdomen**. It lies above and behind the **stomach**, under the ribs on the left side. The spleen is made of spongy **tissue**. It stores blood which can be used if there is a sudden loss of blood in the body. The spleen also makes **white blood cells** which are part of the body's defence, or **immune system.**
The spleen is about the size of an adult's fist.

spondylosis *noun*
Spondylosis is a disorder of the **spine**. It is a kind of weakening and collapse that affects the **vertebrae** and the **discs** between them. Spondylosis in the **neck** causes aching and stiffness. There may be a grating sound when a person turns the neck. Spondylosis in the lower back causes backache and muscle **spasm**.
She had a stiff back due to spondylosis.

spine *noun*

The spine is the backbone of the body. Bones called **vertebrae** sit on top of each other from the neck down to the lower back. Each vertebra is cushioned by strong discs of **cartilage** and held in place with **ligaments** and **muscles**. These allow the spine to twist and bend. The vertebrae are shaped like rings, and surround and protect the **spinal cord**.

The skull rests on the top vertebra in the spine.

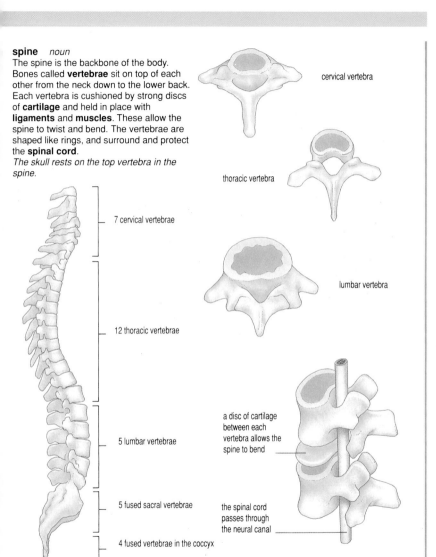

cervical vertebra

thoracic vertebra

lumbar vertebra

7 cervical vertebrae

12 thoracic vertebrae

5 lumbar vertebrae

a disc of cartilage between each vertebra allows the spine to bend

5 fused sacral vertebrae

the spinal cord passes through the neural canal

4 fused vertebrae in the coccyx

spot ► **skin blemish**

sprain *noun*
A sprain is an injury to a **joint**. Sprains happen when the strong strips of **tissue**, or **ligaments**, that hold joints in place are over-stretched or torn. Ankle sprains are often caused by tripping in an awkward way. Sprains are painful and make the tissue around the joint swell. A sprained joint must be rested until it heals.
The skin around a sprain may turn black and blue.

stamina *noun*
Stamina is a measure of **fitness.** Good stamina enables a person to keep up an activity for a long time without getting tired. People build up stamina by taking regular **exercise**. The **heart**, **lungs** and **muscles** become stronger and give the body staying power. Building up stamina is an important part of a healthy lifestyle.
To run a marathon, a person must have a lot of stamina.

stammer ► **stutter**

starch *noun*
Starch is a substance in **food**. Starch is a type of carbohydrate and is found in foods, such as bread, cereals and potatoes. Carbohydrates are one of the important parts of a balanced diet. The starch in food is broken down in the body and used as **energy**.
Iodine can be used to test for the presence of starch in food.

sterile *adjective*
1. Sterile describes a person who is unable to have children. Another word for sterile is infertile. The opposite of sterile is **fertile**.
Certain kinds of disease can make people sterile.
2. Sterile also describes something that is free of **germs** and completely clean.
She applied a sterile bandage to the wound.
sterility *noun*

sterilize *verb*
To sterilize something is to kill all the **germs** living on it. The instruments that doctors use on patients are sterilized. This is to stop germs getting into the body and causing **infection**. People sterilize areas with chemicals such as **antiseptic** and **disinfectant**. Objects can also be sterilized with boiling water or steam, as heat kills germs.
Bandages and plasters have to be sterilized before they are sold.
sterilization *noun*

sternum *noun*
The sternum is a **bone** in the body. It runs down the middle and front of the **chest**. This flat bone is shaped like a dagger. The **collarbones** join the handle of the dagger at the top of the sternum. The first seven pairs of **ribs** curve round and join the blade as it narrows down to a point. The sternum moves in and out with the ribcage as a person breathes. It also helps to protect the **heart** and **lungs** inside.
Another name for the sternum is the breastbone.

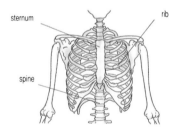

steroid *noun*
A steroid is a **chemical** substance. Many of the **hormones** made by the body are steroids. Steroids can also be made artificially. These kinds of steroid are used as medicines to treat some kinds of illness.
Athletes can be banned from sport if they take steroids.

138

stethoscope   *noun*

A stethoscope is an instrument. **Doctors** and **nurses** use stethoscopes to listen to sounds inside the body. One end is a disc which is pressed against the skin. Two tubes connect the disc to a pair of curved ends that fit into the doctor's ears. Sounds inside the body are picked up by the disc and travel through the tubes to the ears.

The stethoscope felt cold as the doctor placed it on her chest.

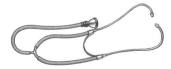

stiffness   *noun*

Stiffness is a **symptom**. It is not being able to bend **joints** or flex **muscles** easily. Stiff muscles and joints are sometimes painful. Exercise and massage are good treatments for all kinds of stiffness.

The old man found it difficult to climb the stairs because of the stiffness in his knees.

stimulate   *verb*

To stimulate is to start or increase action in the body. Most parts of the body react when they are stimulated. Light and sound stimulate eyes and ears. Some **drugs** contain stimulants which activate **organs** in the body.

Caffeine, found in coffee, stimulates the central nervous system.

stimulant   *noun*

stirrup   *noun*

A stirrup is one of three tiny **bones** in the **middle ear**. The other bones are called the **anvil** and the **hammer**. They are all shaped like their names. Together they are called the **ossicles**. The ossicles pick up sound waves and pass them on to the **inner ear**.

The stirrup is the smallest bone in the human body.

stomach   *noun*

A stomach is an **organ** in the upper part of the **abdomen**. Food and liquid enter the stomach from the **esophagus**. Strong **muscles** in the stomach mash up the food. **Acids** in the stomach help to break up the food. Some parts of the food, such as **glucose**, are absorbed into the blood through the stomach walls.

He had eaten so much, his stomach was completely full.

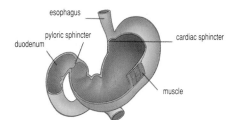

stone   *noun*

A stone is a small, hard lump that forms in the body. Stones are made of **minerals** and **salts** that collect in hollow **organs** or tubes. If they cause a blockage, they need to be removed or dissolved. Common places for stones to form are in the **gall bladder** and **kidneys.**

Some stones can be dissolved within the body by using sound waves.

strength   *noun*

Strength is the power of the body. How strong a person is depends on the strength of their **muscles** and **bones**. Bodies become strong with regular **exercise**. This builds up muscles and **stamina**. Strength develops when a person eats a balanced **diet** of food and has the right amount of **sleep** and rest. Building up strength is an important part of a healthy **lifestyle.**

It took her a long time to regain her strength after her illness.

strong   *adjective*

stress *noun*
Stress is a tension in the mind and body that can cause **illness**. It happens when the body or mind is given too much to do. Stress makes a person worried and anxious. When a person feels stress, they have a faster **heartbeat** and higher **blood pressure**. This can lead to all kinds of other illness. The best treatment for stress is learning how to relax. Some **drugs** also help to relieve stress.
People who hold highly responsible jobs are more likely to suffer from stress.

striated muscle ▶ muscle

stroke *noun*
A stroke is an **illness** that affects the **brain**. It happens when the **blood** supply is cut off from part of the brain. Strokes are caused by a blood **clot** or a burst blood vessel. People recover from mild strokes. More serious strokes cause brain damage which stops a person from moving or speaking. Some strokes lead to a **coma** and death.
The stroke left him paralysed down his left-hand side.

stutter *noun*
A stutter is a **speech** disorder. It can be a symptom of **stress**. A person with a stutter finds it difficult to speak fluently. The beginnings of words are said over and over again before the word comes out. There are gaps between words and the speech is slow. A stutter can be helped with a treatment called speech therapy.
Her stutter became worse when she was embarrassed.

stye *noun*
A stye is a painful, red lump on the **eyelid**. This happens when germs get into the root of an **eyelash** and start an **infection**. **Pus** builds up in the lump. A stye should be bathed with a clean cloth dipped in hot, boiled water. This eases the pain and helps the stye to burst.
Rubbing the eye can make a stye worse.

sucrose *noun*
Sucrose is a kind of **sugar**. It is found in most sweet foods. The sugar used to sprinkle over food is nearly all made of sucrose. The body can use sucrose and turn it into **energy**. Too much sucrose in the **diet** causes bad **teeth** and other problems.
Some sucrose is made from sugar beet.

sugar *noun*
Sugar is a sweet substance in **food**. There are many kinds of sugar in different foods. They can all dissolve in water and be used by the body as **energy**. All sugars are broken down by the body and turned into a blood sugar called **glucose.**
Sugar helps the medicine go down.

surgery *noun*
Surgery is a branch of **medicine** that treats **disease** by carrying out **operations**. A doctor trained in surgery is called a surgeon.
The boy had surgery to remove his appendix.

swallow *verb*
Swallow describes the way food travels to the **stomach** from the **mouth**. After chewing, the **tongue** presses food to the back of the **throat**. A flap of skin called the **epiglottis** shuts off the **trachea** and the food passes into the **esophagus**. This is full of strong muscles which tighten in waves and force balls of food into the stomach.
Laughing while you swallow can cause choking.

sweat   *noun*
Sweat is a salty fluid. It is made by sweat glands in the skin and passed out to the surface through tiny openings, or **pores**. Getting rid of sweat is one of the ways the body keeps cool. It also helps to get rid of urea, a waste product in **blood**. Stale sweat has a strong smell. This is caused by **bacteria** breaking down the mixture of sweat and dead cells on the surface of the **skin.**
His skin was wet with sweat after running in the heat.

sweat gland   *noun*
A sweat gland is a tiny tube that lies coiled under the **skin.** Sweat glands absorb fluid, salts and urea from the **blood.** These waste products are passed out to the surface of skin as sweat. There are sweat glands all over the body. Many of them cluster in the armpits, soles of the feet and palms of the hands. These glands become more active during and after **puberty.**
Sweat glands produce sweat even when a person is cool.

swell   *verb*
To swell is to grow bigger. Swelling in the body is a **symptom**. It may be caused by inflamed or damaged **tissue**. A swelling that forms a lump is called a **tumour**. Swelling also happens when fluid collects in tissue.
His ankle started to swell after his awkward fall during the game.
swelling   *noun*

swelling on a
hand caused
by rheumatism

symptom   *noun*
A symptom is something unusual a person notices about their body. Symptoms can be seen, such as a swelling. Or they may be felt, such as a headache. Symptoms may be inside or on the body, painful or not painful. Doctors need to ask patients about their symptoms to see if they suggest **illness**.
A runny nose is a symptom of a common cold.

syndrome   *noun*
A syndrome is a group of **symptoms**. When these symptoms point to one kind of **illness** they are called a syndrome.
His symptoms amounted to a syndrome of rheumatic fever.

syringe   *noun*
A syringe is a medical instrument. At one end there is a plunger that pushes down into a tube. There is a hollow needle or smaller tube at the other end. Syringes **inject** or draw out fluids, usually through the skin. They can also be used to clean wounds and wash out openings in the body, such as ears.
The nurse used a syringe and needle to give the boy an injection.

tactile   *adjective*
Tactile describes anything to do with the
sense of **touch**. Stroking the skin is a tactile
action.
*The fingers have the most highly developed
tactile sense.*

tapeworm   *noun*
A tapeworm is a flat worm that lives inside
some animals. Humans catch tapeworm by
eating its eggs in undercooked beef, pork or
fish. Hooks on the worm's head cling to the
wall of the **intestine**. Segments of the
worm's body containing eggs fall off and
pass out of the body in the **feces**.
Tapeworms cause **diarrhea**, weight loss and
stomach pain. Worm-killing drugs clear
tapeworm out of the body.
*Tapeworms can grow to over three metres
inside the body.*

tarsals ► **foot**

tartar ► **plaque**

taste ► **tongue**

taste bud   *noun*
Taste buds detect flavours in the **mouth**.
Over 9,000 taste buds surround tiny bumps
on the **tongue**, the back of the mouth and
the **throat**. Taste buds respond to a flavour
when it has dissolved, or become liquid, in
the mouth. **Nerve endings** in the taste buds
pick up flavours and send the message of
taste to the **brain.**
*Taste buds pick up sour, sweet, bitter and
salt tastes.*

tears   *plural noun*
Tears are a watery, salty fluid. They are
made by a gland behind the upper **eyelid**.
The fluid is fed through tubes to the surface
of the **eyeball**. It then drains away through
two small openings in the inner corner of
each **eye**. Tears protect and moisten the eye.
A special **enzyme** in tears kills **germs** and
helps to stop **infection**.
*Blinking helps to spread tears over the
eyeball.*

teenage   *adjective*
Teenage describes a person from 13 to 19
years of age. People pass through **puberty**
and **adolescence** during their early teenage
years. Someone of this age is often called a
teenager.
*She was at school and college all through
her teenage years.*

teeth ► page 144

temperature   *noun*
A temperature is a measure of heat. The
normal temperature of the body is about 37
degrees Celsius. The heat in the body is
controlled by a part of the **brain** called the
hypothalamus. A rise in temperature is a
symptom called a **fever**. A temperature that
drops below normal can lead to an illness
called **hypothermia**. The temperature in the
body is measured with a **thermometer**.
*He had a slight fever, with a temperature of
38 degrees.*

Changes in body temperature (° Celsius)

41	unconsciousness, possible death
	pulse rate rises
37.2	
36	normal body temperature
34	pulse rate falls
30	unconsciousness
25	death

tendon   *noun*
A tendon is a strip of **tissue** in the body.
These strong cords attach **muscles** to
bones. Tendons help muscles to move
bones. A tendon can be strained, or even
torn, by over-use.
*She snapped the tendon in her ankle, and
could not move her foot.*

tension   *noun*
Tension is a strain caused by stretching.
Muscles become tense if they are over-
stretched. This can lead to cramp-like pains.
Tension also describes when the **mind** is
over-stretched by having too much to do at
one time. This type of tension makes a
person worried or anxious. Relaxation is the
treatment for tension in the body or mind.
*A hot bath can give relief for tension in
muscles.*
tense   *adjective*

testis (plural **testes**)   *noun*
A testis is part of a male body. It is one of the
two sex organs contained in a **scrotum**, the
sac of skin that hangs below a **penis**. The
testes are also called the testicles. In an
adult, each testis is about four centimetres
long and shaped like an egg. Testes make
sperm and a male **hormone** called
testosterone. Testosterone helps a boy to
develop into a man.
*Mumps can cause a man's testes to become
inflamed.*

testicle ► testis

tetanus   *noun*
Tetanus is a **disease**. It is caused by **germs**
that live in the soil or animal **feces**. These
germs can enter the body through a cut in
the skin. The first signs of tetanus are a stiff
jaw which makes it difficult to swallow.
Another name for tetanus is lockjaw. The
stiffness spreads down the body and leads to
spasms, **convulsions** and usually **death**.
*Vaccinations can protect people from
tetanus.*

thalamus (plural **thalami**)   *noun*
The thalamus is part of the **brain**. It is one of
two egg-shaped organs that lie above the
hypothalamus. The thalami are **nerve**
centres that pass messages from the eyes,
ears, nose, mouth and skin to the brain.
*Sensory nerves link the mouth to the
thalamus.*

thermogram   *noun*
A thermogram is a way of making a picture of
the inside of the body. Thermograms record
levels of heat, using a special television
camera that is sensitive to heat. This shows
as different colours when printed on paper.
A doctor can see if there is something wrong
with an **organ** by looking at a thermogram.
*A thermogram can reveal blood circulation
problems.*

thermogram
picture of
upper body
and arm

thermometer   *noun*
A thermometer is an instrument that
measures the **temperature** in the body.
Thermometers contain a special liquid,
usually a type of liquid metal called mercury,
in a tube. This liquid rises up the tube when it
is heated. Thermometers can be placed
under the tongue, in the armpit, groin or
anus. The body heat makes the liquid rise.
After a few minutes, this can be measured
against the temperature marked as a scale
on the side of the tube.
*The nurse placed the thermometer in his
mouth to see if he had a fever.*

tooth *noun*

A tooth is a bony shell that grows out of the
gum in the mouth. Each tooth is covered
with a hard layer of **enamel** over a
substance called **dentine**. The soft middle of
a tooth is called the pulp and is filled with
nerves and **blood vessels**. Teeth are used
for cutting and chewing food. Children have
20 milk teeth which fall out and are replaced
by 32 adult, or permanent, teeth.
*Careful brushing helps to keep teeth and
gums healthy.*

A molar tooth

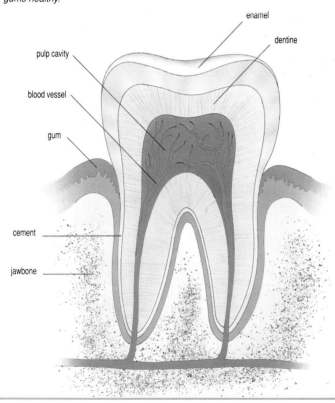

enamel

dentine

pulp cavity

blood vessel

gum

cement

jawbone

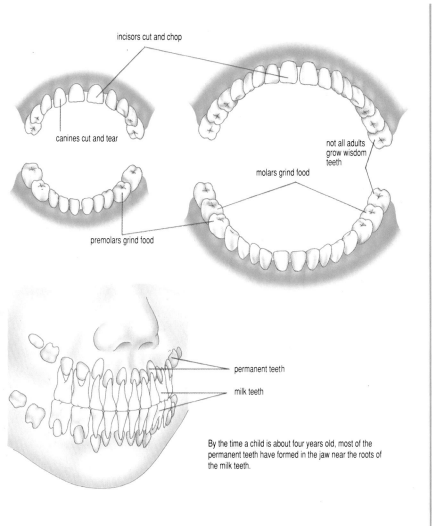

incisors cut and chop

canines cut and tear

not all adults grow wisdom teeth

molars grind food

premolars grind food

permanent teeth

milk teeth

By the time a child is about four years old, most of the permanent teeth have formed in the jaw near the roots of the milk teeth.

thigh *noun*
A thigh is a part of a **leg** between the **knee** and the **hip**. The thighbone, or **femur**, is the longest and one of the strongest bones.
Strenuous exercise gave the weightlifter extremely strong thighs.

think *verb*
To think is to form ideas, or thoughts, in the **mind**. Thinking is part of the **intellect**. Humans have greater powers of thinking than animals.
Students have to spend time thinking in order to pass exams.

thoracic *adjective*
Thoracic describes anything to do with the **chest**.
She learned that her lungs were in her thoracic cavity.

thorax ► **chest**

threadworm *noun*
A threadworm is a small worm that lives in the **large intestine**. The female worm crawls down the rectum to lay her eggs around the **anus**. This causes severe itching and can lead to **infection**.
The doctor used drugs to treat an infection caused by threadworm.

throat *noun*
The throat is the front part of the **neck**. It starts at the back of the **mouth** in the **pharynx**. The **larynx** and openings to the **esophagus** and **trachea** are all in the throat.
Large veins and arteries pass through the throat.

thrombosis *noun*
A thrombosis is a blood **clot** that forms in an artery or vein. A blood clot that breaks loose and travels around a **blood vessel** is called an **embolus**. Thrombosis prevents blood flowing properly by causing a blockage.
A coronary thrombosis results in damage to the heart muscle.

thumb *noun*
A thumb is the first **finger** on each **hand**. Thumbs are shorter and fatter than other fingers and only have two bones, whereas the other four fingers each have three. The base of each thumb has a **joint** unlike any other in the body. These unusual joints are called saddle joints. They allow thumbs to move from side to side and backwards and forwards.
Because the human thumb can push against the fingers it is called an opposable thumb.

thymus *noun*
A thymus is an **organ** in the **chest**. It lies above and in front of the **heart**. The thymus makes **white blood cells** which help to protect the body from illness. Babies are born with a large thymus, which gradually doubles in size during childhood. After **puberty** the thymus starts to shrink and is less active.
The thymus is an organ which is part of the lymphatic system.

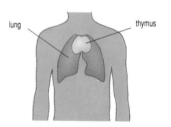

lung · thymus

thyroid gland *noun*
A thyroid gland is an **organ** in the neck. It lies around the front and sides of the **trachea** just below the **larynx**. The thyroid gland releases **hormones** into the blood which affect all the cells in the body. These hormones control how quickly the body burns up energy. This is known as the **metabolic rate** of the body.
Hormones from the thyroid gland are important for a child's mental development.

tibia   *noun*
A tibia is a **bone** in the leg. The tibia is the longer of the two bones in the lower leg. The thinner outer bone is called the **fibula**. The top of the tibia forms part of the **knee**. The lower end forms part of the **ankle**.
The tibia is also called the shinbone.

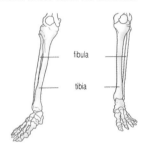

fibula

tibia

tic   *noun*
A tic is a rapid twitching of a muscle which a person cannot control. Tics often happen in the muscles of the face.
The boy's blinking eye was an annoying tic.

tick   *noun*
A tick is a small animal that sucks **blood**. Some ticks have soft bodies and others have hard, shell-like backs. Ticks bury their heads in the **skin** of a human or animal to feed. Some kinds of tick carry **disease**. A tick should be dabbed with alcohol before carefully taking it out with tweezers. Pulling at it may leave the head behind and cause **infection**.
Ticks swell up as they feed on blood.

tinnitus   *noun*
Tinnitus is a sound heard inside the **ear**. It may be a ringing, buzzing or hissing noise. The sounds heard do not come from outside the ear. Tinnitus may be caused by wax in the ear, or head or ear injuries. Tinnitus may be linked with **deafness**.
Tinnitus kept him awake at night.

tissue   *noun*
Tissue is the fabric of the body. Clumps of **cells** of the same kind are arranged in layers to form tissue. Different types of cell are grouped together to make **skin** and every other **organ** in the body. Living tissue is fed with **oxygen** and **nutrients** from blood. Dead tissue, such as hair, has no blood supply.
The tissue round her sprained wrist was swollen and puffy.

toe   *noun*
A toe is found at the end of a **foot**. There are five toes ending in five toenails on each foot. The big toe has only two bones, like the **thumb** of each hand. Each of the other toes has three bones. Toe bones, or **phalanges**, are joined together by **joints** that act like hinges and allow them to bend. Toes help to spread the weight of the body evenly. They also help the body to **balance** on the feet.
When he broke his toe, he found it very difficult to walk.

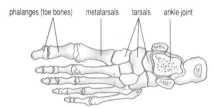

phalanges (toe bones)   metatarsals   tarsals   ankle joint

tongue   ► page 148

tonsil   *noun*
A tonsil is a lump of spongy **tissue** at the back of the **mouth**. There are two main tonsils that lie either side of the entrance to the **throat**. Tonsils help to prevent **germs** getting into the throat. Sometimes very large tonsils make breathing difficult. When this happens the tonsils can be removed by a simple operation.
Children have bigger tonsils than adults.

tongue   *noun*

The tongue is a muscular **organ** in the **mouth**. It is attached to the back and the floor of the mouth. The tip of the tongue is free to move all around and out of the mouth. Tongues are covered with moist, bumpy skin. Thousands of **taste buds** on the surface of the tongue pick up flavours in food and drink. Tongues are used to push food into the **pharynx**. Movements of the tongue also help to make the sounds of **speech**.

The number of taste buds in the tongue decreases as a person gets older.

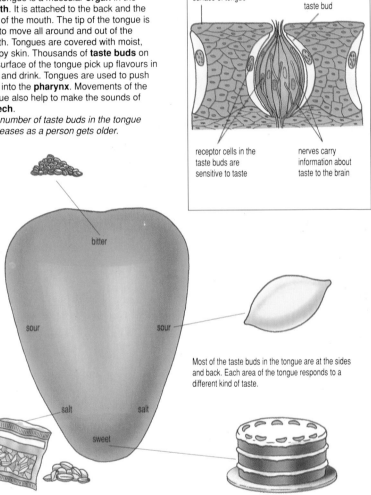

surface of tongue

taste bud

receptor cells in the taste buds are sensitive to taste

nerves carry information about taste to the brain

bitter

sour   sour

salt   salt

sweet

Most of the taste buds in the tongue are at the sides and back. Each area of the tongue responds to a different kind of taste.

tonsillitis   *noun*
Tonsillitis is a common childhood **illness**.
It happens when the tonsils are infected by
bacteria. The **throat** becomes sore and
swollen. There may be a headache and
fever. Tonsillitis is treated with **antibiotic
drugs** that kill the germs causing the
infection. If tonsillitis keeps coming back, the
tonsils can be surgically removed to cure the
problem.
*Her tonsillitis was so bad that she could
hardly swallow.*

tooth ▶ **teeth**

toothache   *noun*
Toothache is a pain in a tooth, or around a
tooth. It is usually caused by decay. Decay
eats away the hard, outer layer of a tooth.
When the **nerve** inside is exposed, a person
feels toothache. Most toothache is cured by
filling or removing the tooth.
He took a painkiller to relieve his toothache.

touch ▶ page 150

toxin   *noun*
A toxin is a poison. Toxins are made in the
cells of animals and plants. They enter the
body through **germs** carrying disease. The
body fights toxins by producing **antitoxins** in
the blood. Specially treated toxins are
sometimes given to people as **vaccines**.
*The venom of a poisonous snake is a kind of
toxin.*
toxic   *adjective*

trachea   *noun*
The trachea is a tube in the body. The
trachea is the air passage from the **larynx** in
the throat to the lungs. It runs from the **throat**
down to just above the **heart** in the middle of
the **chest**. Here it divides into two tubes, or
bronchi, that lead to each **lung**. A flap of
skin called the **epiglottis** closes the top of
the trachea when a person is drinking.
*The trachea of an adult is about 13
centimetres long.*

transfusion   *noun*
A transfusion is a way of giving a person
blood. Fresh blood from a healthy person is
fed through a hollow needle into a **vein**.
Transfusions are given to people who do not
have enough blood because of an accident
or disease. The blood they are given must be
the same type, or **blood group**, as their own
blood. **Plasma** can also be given by
transfusion.
*Blood given by donors for transfusion can be
stored in a blood bank until it is needed.*

transplant   *noun*
A transplant is a way of taking an **organ** or
part of the body from one person, or donor,
and putting it into another person, or
recipient. Some organs, such as the heart,
are taken from people who have recently
died. Sometimes, a living donor gives part of
the body that they can live without, such as a
kidney. For a successful transplant, the body
tissue of the donor and the recipient must be
of a similar type. In this way, the recipient's
body does not reject the transplanted organ.
*Her sight improved dramatically after the
corneal transplant.*

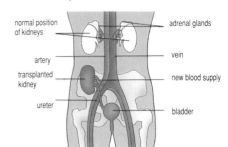

normal position of kidneys | adrenal glands
artery | vein
transplanted kidney | new blood supply
ureter | bladder

trauma   *noun*
Trauma is a medical word for an injury or
wound. It can also be emotional **shock**
caused by an accident or great unhappiness.
*She suffered trauma after the horrific train
crash.*

touch *noun*

Touch is a **sense**. It is being able to feel something when the **skin** comes in contact with it. **Sensory nerves** in the skin send messages to the **brain**. Some parts of the body, such as fingertips and lips, are very sensitive to touch. This is because the skin that covers them has more nerve endings than other parts of the body.

Blind people often have a specially sensitive sense of touch.

There are different kinds of nerve endings in the skin.

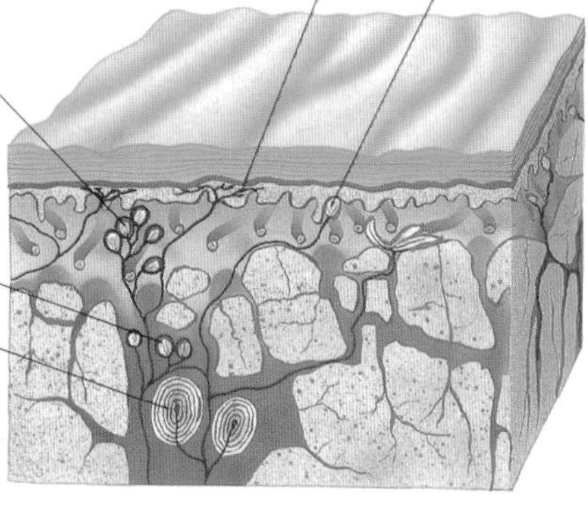

sensitive to vibration and pressure

sensitive to texture and pressure

feel cold

feel heat

free nerve endings feel pain

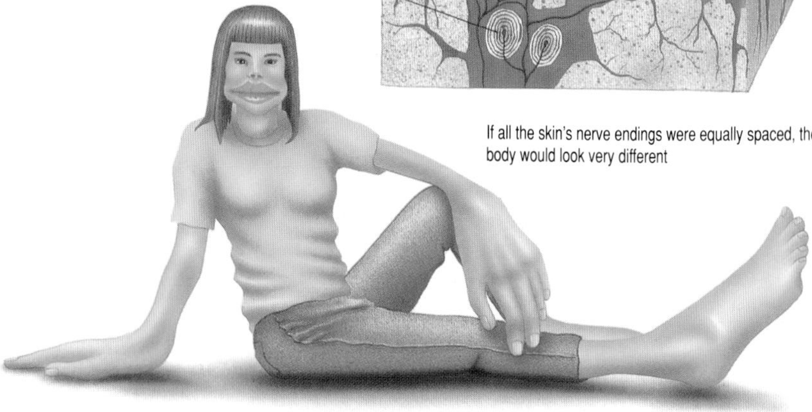

If all the skin's nerve endings were equally spaced, the body would look very different

treatment   *noun*
Treatment is the care of **illness**. Giving drugs
and other kinds of **medicine** is often part of
any treatment. Resting is another way of
treating illness. The treatment for a small cut
is to clean it, put on antiseptic and a plaster.
Some treatment, such as an **operation**, is
carried out in a **hospital**.
The doctor said the best treatment was to
rest and drink plenty of water.

triceps   *noun*
A triceps is a **muscle**. The triceps lies along
the back of the upper **arm** and helps to move
the forearm. This strong muscle works with
the **biceps** muscle on the front of the upper
arm. The triceps tightens to move the arm
into a straight line. The biceps tightens to
bend the arm at the **elbow**. When one
muscle is tight and short the other is long
and relaxed.
The triceps works with the biceps to move
the forearm.

trichinosis   *noun*
Trichinosis is an **illness.** It is caused by tiny,
round worms that live inside the **stomach**,
intestine and **muscles**. These kinds of
worm get into the body when a person eats
under-cooked pork. Diarrhea and vomiting
are the first **symptoms**. Two weeks later the
eyelids swell and the skin covering the
eyeball becomes red and inflamed. Aching
muscles, fever and weakness follow.
Trichinosis can be prevented by cooking pork
thoroughly.

tuberculosis   *noun*
Tuberculosis is an infectious **disease** caused
by **bacteria**. It attacks the **lungs** and may
then spread to other parts of the body. There
are fevers, night sweats and weight loss.
A person coughs up blood as the lungs
become more damaged. **Antibiotic drugs**
cure tuberculosis. There is also a **vaccine**
which protects people from catching this
disease.
Tuberculosis used to be called consumption.

tumour   *noun*
A tumour is a **swelling** in the body. This can
be any kind of lump, such as an **abscess** or
cyst. A tumour may be a growth of **tissue**
which is harmless, or **benign**. A tumour may
be **malignant** and a symptom of **cancer**.
A malignant tumour needs to be removed
before it spreads to other parts of the body.
She was relieved that her tumour was
benign.

twins   *plural noun*
Twins are two **babies** who grow in the
uterus at the same time. The babies are
born one after the other. Identical twins
happen when a fertilized female **ovum**
divides into two. The two ova develop into
two babies of the same sex who look alike.
Non-identical twins happen when two ova
are fertilized by two sperm. These babies
may be different sexes and look no more
alike than any other brother or sister.
It was difficult to tell the identical twins apart.

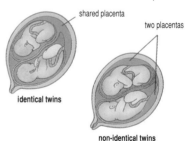

shared placenta

two placentas

identical twins

non-identical twins

typhoid fever   *noun*
Typhoid fever is a kind of food poisoning
caused by **bacteria** in the **intestine**. This
causes a high fever, constipation and a **rash**
over the chest and abdomen. Diarrhea and
more serious illness may follow. Typhoid is
carried in food and water contaminated by
feces or urine. **Antibiotic drugs** are used to
treat this disease.
People who have recovered from typhoid
may still carry and pass on the infection.

typhus *noun*
Typhus is a general term for a group of
diseases. These are carried by rat fleas,
lice, ticks and mites in areas where the
standard of **hygiene** is low. Two weeks after
a bite from one of these blood-sucking
insects, a person develops a severe
headache and a high fever. A pink **rash**
breaks out over most of the body. **Vaccines**
can help protect people from some forms of
typhus.
*The spread of typhus can be controlled by
killing the insects that carry the disease.*

ulcer *noun*
An ulcer is an open sore. Ulcers may
develop on the inside or on the outside of the
body. There are many different kinds of ulcer.
Some are mild and heal quickly, such as
those caused by a burn or by a graze. Others
are **symptoms** of a more serious illness.
Ulcers that develop inside the **stomach**,
such as a peptic ulcer, may take a long time
to heal.
*Her mouth ulcer hurt when she ate the
tomato.*

ulna *noun*
An ulna is a **bone** in the lower **arm**. It is the
bone that runs from the back of the elbow to
the **wrist**. The top end of the ulna joins the
bone of the upper arm called the **humerus** to
form part of the **elbow**. The lower end forms
part of the wrist on the side of the little finger
of the **hand**. The other bone in the arm is the
radius.
*The ulna twists over the radius to allow the
wrist to turn.*

ultrasound *noun*
Ultrasound is sound waves. The human **ear**
is unable to hear ultrasound as it vibrates
very fast through the air. Ultrasound can be
passed into the body in a thin beam. Solid
objects, such as **bone**, reflect the sound
waves. Less dense **tissue** and **organs**
absorb them. The echo from ultrasound can
be used to build up a picture of the inside of
the body. It is often used to examine the
fetus in early **pregnancy**.
*The ultrasound picture showed that the
baby's spine was normal.*

umbilical cord   *noun*
An umbilical cord is a coil of **blood vessels**
surrounded by **skin**. It joins a mother to her
unborn baby. One end of the cord is attached
to the baby in the middle of the **abdomen**.
The other end joins the **placenta** on the wall
of the **uterus**. **Oxygen** and **nutrients** pass
through the placenta and cord to the baby.
After **childbirth**, the cord is cut. The stump
shrivels and falls off, leaving a scar called a
navel.
*The umbilical cord is usually about
60 centimetres long.*

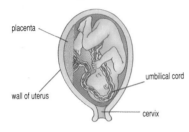

placenta

umbilical cord

wall of uterus

cervix

unconscious   *adjective*
To be unconscious is to be not awake.
A person may be fully unconscious after a
head injury. This looks like a deep **sleep.**
Or a person may be semi-conscious, or half
awake, such as when a person is recovering
from a fit or a **faint**. Someone who is deeply
unconscious through accident or illness is
said to be in a **coma.**
*She was knocked unconscious when the
brick hit her head.*

ureter   *noun*
A ureter is one of two thin tubes in the body.
The top end of each ureter is attached to a
kidney. The lower ends join the **bladder**.
Urine drains into the ureters from the
kidneys and is passed into the bladder.
Muscles in the ureters help to squeeze the
urine down.
*The muscles in the ureters contract and relax
about three times a minute.*

urethra   *noun*
A urethra is a tube in the body. It is the
narrow tube from the bladder through which
urine is passed out of the body. The urethra
in a male travels down the **penis** into an
opening at the end. The male urethra also
acts as a passage for **semen**. The opening
of the female urethra lies between the
vagina and the **clitoris**.
*A man's urethra is much longer than a
woman's.*

urine   *noun*
Urine is a yellow fluid made in the body. It is
mostly water mixed with salts and **waste
products**, such as urea, filtered from the
blood. Urine is made by the **kidneys** and
carried to the **bladder** by the **ureters**, where
it is stored. It is then passed out of the body
as a waste product. Normally, urine is a
yellow colour.
*Eating beetroot can change the colour of
urine from yellow to red.*

urine test   *noun*
A urine test looks for substances in the
urine. A small amount of urine is mixed with
chemicals. This can show if there is anything
unusual in the urine, such as **blood** or **pus**.
Urine tests can tell a doctor if a person has a
certain kind of illness, such as **diabetes**. One
of the signs of diabetes is sugar in the urine.
A urine test can also show if a woman is
pregnant.
*The urine test showed that she had a small
amount of glucose in her urine.*

uterus   *noun*
A uterus is an **organ** in a female. This
strong, pear-shaped bag lies in the **pelvis** at
the top of the **vagina.** Unborn babies grow in
the uterus, which becomes bigger during
pregnancy. Every month after **puberty** the
uterus grows a thick lining in case a woman
becomes pregnant. If no **ovum** is fertilized,
the lining is shed and passed out of the body
during **menstruation**.
The uterus is also called the womb.

uvula   *noun*

An uvula is a small, soft lump of **tissue**. It hangs from the soft **palate** at the back of the mouth. The uvula dangles over the entrance to the **throat**. It can be seen when the mouth is opened wide and the **tongue** held down.
The singer opened his mouth so wide, everyone could see his uvula.

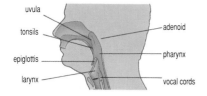

uvula
tonsils
epiglottis
larynx
adenoid
pharynx
vocal cords

vaccination   *noun*

A vaccination is a dose of **medicine** made from a **vaccine**. Vaccinations are given through the mouth or a scratch on the skin. They are most often injected through the skin. Vaccinations prevent all kinds of **disease**, such as **measles** and **diphtheria**. Children are given vaccinations to protect them from catching these diseases.
The doctor recommended he should have a tetanus vaccination every 10 years.

vaccine   *noun*

A vaccine is a **medicine** which stops people catching a **disease**. It is made from the **germs** that cause the disease. These may be living or dead. Sometimes, a vaccine is made from poisons produced by germs. Giving someone a vaccine makes the body's defence, or **immune system**, attack the germs. The person then has immunity, or protection from ever catching this disease.
The poliomyelitis vaccine is given as drops into a baby's mouth.

vagina   *noun*

A vagina is a female sex organ. It is the passage between the **vulva** and the **uterus** inside the body. The opening of the vagina lies between the **urethra** and the **anus** at the top of the legs. The soft, elastic walls of the vagina fold together with the action of **muscles**. They are covered with a moist **membrane**. This helps to keep the vagina clean and free from germs.
After puberty, the vagina is important in menstruation, sexual intercourse and childbirth.

valve *noun*

A valve is a flap of **tissue** in a tube. Valves prevent a liquid from flowing backwards. Valves in the **veins** and **heart** make sure **blood** flows in the right direction. Another valve in the body lies between the small and large **intestine**. This stops liquid food from moving backwards.

The semilunar valve in the heart has three flaps shaped like half moons.

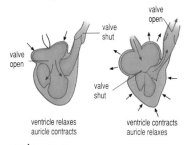

ventricle relaxes
auricle contracts

ventricle contracts
auricle relaxes

vein *noun*

A vein is a tube inside the body. There is a network of veins inside the body. They all carry **blood** from the body tissues back to the **heart**. Tiny veins called **capillaries** pass the blood to larger veins called venules. These join bigger veins which connect up to two huge veins that empty blood into the heart. The blood flow in a vein is helped by muscles which surround it. The blood can only flow one way because many of the veins are fitted with **valves**.

Veins have thinner walls than arteries.

ventricle *noun*

A ventricle is a space in the **heart**. It is one of the two lower **cavities**, or chambers which pump **blood** into the **arteries** during **circulation**. The right ventricle pumps blood to the **lungs**. The left ventricle pumps blood to the **arteries**. The ventricles receive blood from the **atria**, the two upper chambers of the heart.

The ventricles of the heart are stronger than the atria.

verruca *noun*

A verruca is another name for a **wart**, which is a skin infection due to a **virus**.

The verruca on the sole of his foot was quite painful.

vertebra (plural **vertebrae**) *noun*

A vertebra is a **bone** in the **spine**. There are 33 vertebrae in the backbone from the **neck** down to the **coccyx**. The smallest vertebrae are in the neck and they get bigger further down the spine. The upper 24 vertebrae are cushioned between strong **discs** of **cartilage**. These flexible, ring-shaped bones surround and protect the **spinal cord**.

The lowest four vertebrae are fused together to form the coccyx.

vertebrate *adjective*

Vertebrate describes an animal that has a **backbone**.

Rabbits and snakes are vertebrate animals.

vertigo *noun*

Vertigo is a disorder of **balance**. It is the feeling of spinning around when a person stays still. Sometimes, the objects outside the body seem to be spinning around. People with vertigo often feel sick and **vomit**. Vertigo may be caused by a blockage or infection in the **ear**. Alcohol and **drugs** sometimes cause vertigo. It may also be a sign of other illness. A person suffering from vertigo should lie down with the eyes closed.

Climbing to the top of the tower gave him a bad attack of vertigo.

virus *noun*

A virus is a tiny **germ**. Viruses are much smaller than **bacteria** and can only be seen with a powerful microscope. They only survive in living **tissue**. They infect **cells** and alter them. Different viruses attack different kinds of cells in the body. Many diseases, such as measles and influenza, are caused by viruses. **Vaccines** protect people from some of these diseases.

The common cold is caused by a virus.

voice *noun*

A voice is a sound made by the **vocal cords**. The two vocal cords are part of the **larynx** in the throat. They are folds of skin with a gap between them called the **glottis**. A voice is heard when air from the **lungs** pushes past the vocal cords, making them vibrate. Voices enable people to speak, sing and make noises.

After puberty, male voices sound lower than female voices.

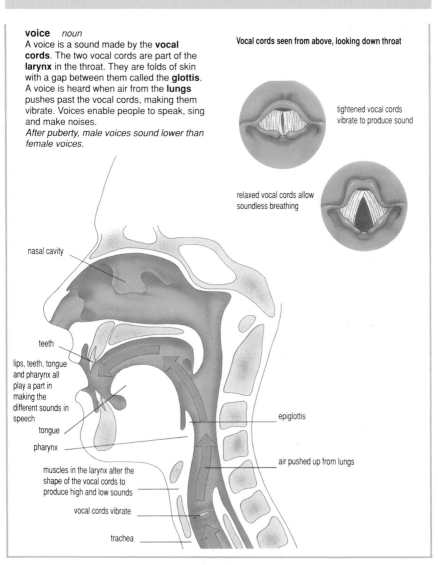

Vocal cords seen from above, looking down throat

tightened vocal cords vibrate to produce sound

relaxed vocal cords allow soundless breathing

nasal cavity

teeth

lips, teeth, tongue and pharynx all play a part in making the different sounds in speech

tongue

pharynx

muscles in the larynx alter the shape of the vocal cords to produce high and low sounds

vocal cords vibrate

trachea

epiglottis

air pushed up from lungs

vision ► **sight**

vitamin *noun*
A vitamin is a chemical substance. The body
needs small amounts of all the different
vitamins to grow and stay healthy. Some
vitamins are known by letters of the alphabet.
These are A, B, C, D, E and K. Most vitamins
are found in different kinds of **food**. Eating a
balanced **diet** of food usually gives a person
all the vitamins they need.
*Oranges and blackcurrants contain a lot of
vitamin C.*

vitiligo *noun*
Vitiligo is a **skin** disorder. Small patches of
pale-coloured skin appear on the body
because of a lack of **melanin**.
Sitting in the sun helped to cure her vitiligo.

vitreous humour ► **eyeball**

vocal cord *noun*
A vocal cord is part of the **larynx** in the
throat. The two vocal cords are folds of skin
with a narrow space between them called the
glottis. Air passes between the vocal cords
when a person breathes. When the cords
move together, the glottis becomes a tiny slit.
This makes the vocal cords vibrate as air
passes through.
*Vibrating vocal cords make the sounds of
speech, song and all other noises.*

voice ► page 156

voice box ► **larynx**

vomit *verb*
To vomit is to throw up **food** from the
stomach back out of the mouth. It is often
called being sick. The vomited food is usually
sour smelling and semi-liquid. Vomiting is a
common **symptom** of many kinds of mild
and serious illness. Food poisoning and
gastroenteritis cause vomiting.
*He vomited after he had eaten the infected
shellfish.*

vulva *noun*
The vulva is the outer sex organ of a female.
It is the part that lies between the legs. The
soft, padded triangle at the top of the vulva is
called the mons pubis. Further down the
vulva are folds of skin called **labia**. These
surround the **clitoris** and the openings of the
urethra and **vagina**. The entrance to the
vagina is sometimes partly covered by a fold
of skin called the hymen.
*The vulva grows hair and changes shape
and colour during puberty.*

wart *noun*
A wart is a knobbly growth on the **skin**. Warts are caused by a **virus**, and they vary in size, shape and colour. Warts are not usually painful, and they disappear in their own time if left alone. But if a plantar wart, or **verruca**, on the sole of the foot is painful, it can be removed.
Some people believe that warts can be charmed away.

waste product *noun*
A waste product is a substance not needed by the body. Waste products are either made in the body or taken in by the body. Some **organs**, such as the **kidney**, collect and destroy waste products. Many waste products are passed out of the body as **feces** and **urine**. Breathing out gets rid of a waste product in the air called **carbon dioxide**. **Sweating** helps the body to get rid of a waste product in the **blood** called urea.
If waste products are not removed from the body, a person can become ill.

wax *noun*
Wax is a sticky, yellow substance made in the **ear**. Wax helps to protect the ear from **infection**. Wax sometimes builds up and forms a hard plug over the **eardrum** This may cause **deafness**. Hard wax can be softened and removed by a doctor.
His hearing improved after the doctor removed the hard wax from his ears.

white blood cell ► **blood**

white matter ► **brain**

whooping cough *noun*
Whooping cough is a childhood **disease** caused by **bacteria**. Whooping cough is picked up by the coughing of an infected person. Violent, short coughs are followed by a deep intake of breath which makes the 'whoop' sound. Whooping cough causes vomiting, and can lead to **pneumonia** and **brain** damage. It may take several months to recover from the disease.
Doctors recommend that all children should be vaccinated against whooping cough.

windpipe ► **trachea**

wisdom tooth ► **tooth**

womb ► **uterus**

worm *noun*
A worm is an animal with a long, soft body and no legs. Different kinds of worm, called parasites, live inside humans. **Tapeworm** and **hookworm** are two kinds of worm parasites. Children often catch tiny threadworms that can be seen wriggling in their **feces**. These are soon cleared out of the body with worm-killing medicines.
The worms made the boy itch round the anus.

wrist *noun*
A wrist is the part of the body that joins the **hand** to the **arm**. The wrist has eight small **bones** in two rows. The bones of the wrist meet the bones of the forearm and hand to form the wrist.
More than 20 tendons link the wrist, the arm and the hand.

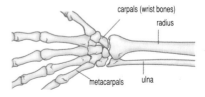

carpals (wrist bones)
radius
metacarpals   ulna

X-chromosome *noun*

An X-chromosome is part of a **cell**. It is one
of the two **chromosomes** that make a
person **male** or **female**. The other one
is called a **Y-chromosome**. Two
X-chromosomes in a fertilized **ovum** make
a female.
Some diseases are caused by defects of the
X-chromosome.

X-ray *noun*

An X-ray is a photograph of the inside of the
body. X-ray pictures are black and white
images of **bones** and **organs**. In an X-ray,
the solid areas look light and the spaces in
between look dark. Taking X-rays is a branch
of medicine called **radiography**.
The X-ray showed a broken tibia.

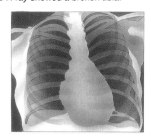

yaws *noun*

Yaws is a **disease** found in tropical
countries. It is caused by a **bacterium**. Soft
swellings appear on the lips, elbows, knees
and buttocks. They break open and heal
slowly. The sores may damage tissue and
bones underneath the skin. The disease may
disappear for several years before coming
back. Yaws can be cured with **antibiotic**
drugs.
The germ that causes yaws can only infect
another person through broken skin.

Y-chromosome *noun*

A Y-chromosome is part of a male cell. It is
one of the two **chromosomes** that make a
person **male** or **female**. The other one is
called an X-chromosome. If a **sperm**
carrying a Y-chromosome meets an
X-chromosome in a female **ovum**, the baby
will be a boy. A sperm decides the sex of a
child at **conception**.
A Y-chromosome looks different from all the
other chromosomes in a cell.

yellow fever *noun*

Yellow fever is a **disease** of tropical areas
that can only be caught once. It is caused by
a **virus**, passed on by a mosquito bite. A
high **fever** follows shaking chills, headache
and pains inside the body. Many people
recover after these **symptoms**. Some people
become more ill as the **virus** attacks the
kidneys and **liver**. This turns the skin yellow,
giving the disease its name. A small number
of people fall into a coma and die.
Yellow fever vaccine gives people immunity
from the disease for 10 years.

yoga *noun*
Yoga is a way of keeping fit and **healthy**.
Yoga exercises make the body strong and
supple, or able to bend and stretch easily. In
yoga, the exercises are done with a special
way of breathing. Yoga helps people feel
calm and relaxed. It is very good for stress.
People in India have been doing yoga for
about 2,000 years.